Pearson Australia
(a division of Pearson Australia Group Pty Ltd)
459–471 Church Street, Level 1, Building B, Richmond, Victoria 3121
PO Box 23360, Melbourne, Victoria 8012
www.pearson.com.au

First published 2016 by Pearson Australia
2026 2025 2024 2023
10 9 8 7 6 5 4 3 2 1

Publisher: Alicia Brown
Development Editor: Shirley Melissas
Project Manager: Shelly Wang
Production Manager: Elizabeth Gosman
Editors: Deborah Lombardo, Aptara
Proof reader: Vicky Chadfield
Designer: Anne Donald
Copyright & Pictures Editor: Samantha Russell-Tulip
Desktop Operators: David Doyle, Aptara
Illustrator: DiacriTech
Printed in Malaysia (CTP-VVP)

ISBN 9781488615078

Pearson Australia Group Pty Ltd ABN 40 004 245 943

Acknowledgements
We would like to thank the following for permission to reproduce copyright material. The following abbreviations are used in this list: t = top, b = bottom, l = left, r = right, c = centre.

Cover image: Science Photo Library/Marek Mis

123RF: pp. 49, 57cl, 140; Kadriya Gatina, p. 168; grazvydas, p. 57cr; Pavel Sazonov, p. 7.

AFP: Extract from Marlowe Hood, "Scientists reveal how snakes 'see' at night", 15/03/2010, p. 63.

Alamy Stock Photo: Agencja Fotograficzna Caro, p. 107b; Jonathan Need, p. 132.

Australian Microscopy & Microanalysis Research Facility (AMMRF): Extract from 'Indigenous medicine meets biomaterials development', AMMRF News, Vol. 5, 2009 (p. 1 in original), p. 105-06.

Bertram Lobert: p. 137b.

Capella Observatory: www.capella-observatory.com/Stefan Binnewies and Josef Pöpsel, p. 66.

CSIRO Publishing: CSIRO Entomology, p. 135.

Fotolia: Heike Falkenberg, p. 18; mahony, p. 5b.

Getty Images: Photo Researchers, p. 107t; Frank Schwere, p. 20.

Microsoft Corporation: Used with permission from Microsoft, pp. xi, xii, xiii.

Science Photo Library: Marek Mis, p. i; Alfred Pasieka, p. 111.

Shutterstock: AboliC, p. xiii; dezign80, p. 57tl; dubassy, p. 57tc; FloridaStock, p. 121; Four Oaks, p. 137t; Alena Hovorkova, p. 160; Levent Konuk, p. 57c; Ye Liew, p. 163; Paul Maguire, p. 57br; MrGarry, p. 57tr; tobkatrina, p. 23.

Sozaijiten: p. 5t.

Thinkstock: Brand X Pictures, p. 63tr.

Every effort has been made to trace and acknowledge copyright. However, should any infringement have occurred, the publishers tender their apologies and invite copyright owners to contact them.

PEARSON Australian Curriculum Writing and Development Team

Anna Bennett
Differentiation consultant, Victoria

Ian Bentley
Former Head of Science, VCE exam and trial exam writer, STEM investigation developer, Victoria

Christina Bliss
Science and senior Biology teacher, VCE assessor, Biology author and question writer, Victoria

Donna Chapman
Science laboratory technician, safety consultant, Victoria

Dr Warrick Clarke
Curriculum writer, Science communicator and Australian Post-Doctoral Research Fellow at UNSW, author, New South Wales

Jacinta Devlin
Science and senior Physics teacher, author, Victoria

Julia Ferguson
Education Officer Earth Science Western Australia, author and reviewer, Western Australia

Bob Hoogendoorn
VCAA exam assessor in Chemistry, Former senior Chemistry teacher exam-style question author, Victoria

Penny Lee
Science laboratory technician, safety consultant, Victoria

Louise Lennard
Head of Science, former industrial scientist, author and STEM investigation developer, Victoria

Greg Linstead
Former Head of Science, Education Department of WA curriculum writer, Deputy Chief Examiner, STAWA President, author, Western Australia

Bryony Lowe
Director of Numeracy Improvement. Former Head of Science, Region Teaching & Learning Coach, author and reviewer, Victoria

David Madden
Science Learning Area Manager at QCAA, former Head of Science, author and reviewer, Queensland

Fran Maher
Bioscience educator, Science teacher, former Bioscience researcher author, Victoria

Rochelle Manners
Science and Mathematics teacher, Teacher Companion coordinating author, Queensland

Shirley Melissas
Teacher librarian and author, Development editor, Victoria

Tamsin Moore
Science and senior Psychology teacher, author, Western Australia

Natalie Nejad
Head of Science, Science and Mathematics teacher, STEM investigation developer, Victoria

Malcolm Parsons
Education consultant, former teacher, author, Victoria

Greg Rickard
Teacher, former Head of Science, author, Victoria

Lana Salfinger
Teacher, Head of Science, IB Workshop leader in MYP Sciences author, Western Australia

Maggie Spenceley
Former teacher, curriculum writer Queensland Studies Authority, author, Queensland

Jim Sturgiss
Teacher, former coordinating analyst (NSW Department of Education) reporting NAPLAN, senior test designer for Essential Secondary Science Assessment (ESSA), Thinking Scientifically questions author, New South Wales

Craig Tilley
STAV trial exam coordinator, VCAA exam assessor, former senior Physics teacher, author and exam-style question writer, Victoria

Jo Watkins
Chief Executive Officer at Earth Science Western Australia, author and reviewer, Western Australia

Dr Trish Weekes
Science literacy consultant, New South Wales

Rebecca Wood
Science educator and tutor, author, Victoria

Contents

Contents *continued*

Access the worksheets for the Step-up chapter in Psychology on your eBook. Answers are available in the teacher version of the eBook and on Productlink

Pearson Science 2nd edition Activity Book

An intuitive, self-paced approach to science education, which ensures every student has opportunities to practise, apply and extend their learning through a range of supportive and challenging activities.

Pearson Science 2nd edition has been updated to fully address all strands of the new Australian Curriculum: Science, which has been adopted throughout the nation. This edition also captures the coverage of Science curricula in states such as Victoria, which have tailored the Australian Curriculum slightly for their students.

The *Pearson Science 2nd edition* features a more explicit coverage of the curriculum. The activities enable flexibility in the approach to teaching and learning. There are opportunities for extension as well as reinforcement of key concepts and knowledge. Students are also guided in self-reflection at the end of each topic.

Explicit scaffolding makes learning objectives clear and includes regular opportunities for reflection and self-evaluation.

In this edition, we provide a structured approach that integrates a seamless, intuitive and research-based learning hence **differentiating** the course for every student.

The Activity Book also provides richer application opportunities to take the Student Book content further with explicit coverage of Inquiry Skills, Science as a Human Endeavour and Science Understanding.

The diverse offering of worksheets allows students to be challenged at their level. Students have the flexibility to be self-paced and this new edition comes with the advantage of each worksheet being self-contained.

Be guided

A new handy **Toolkit** at the beginning of the Activity Book has been created to build skills in the key areas of practical investigations, research, thinking, organising, collecting and presenting. Each skill developed in the toolkit is directly relevant to applications in questions, investigations and research activities throughout the student and activity books. A toolkit spread provides guides and checklists alongside models and exemplars.

Be supported

Vocabulary boxes provide definitions for key terms, within the relevant context of the task. **Hints** help students get started on a worksheet and provide support in overcoming a barrier.

Be reflective

The **Thinking about my learning** feature provides the opportunity for self-reflection and self-assessment. It encourages students to look ahead to how they can continue to improve and assists in highlighting focus areas for skill and knowledge development.

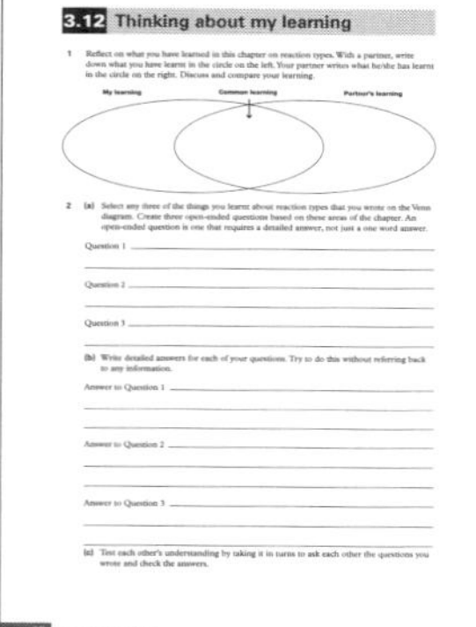
3.12 Thinking about my learning

Be ready

A **Knowledge preview** at the beginning of every chapter, activates prior knowledge relevant to the topic, providing an opportunity for students to show what they currently know. This handy tool supports teachers in assessing students' prior knowledge.

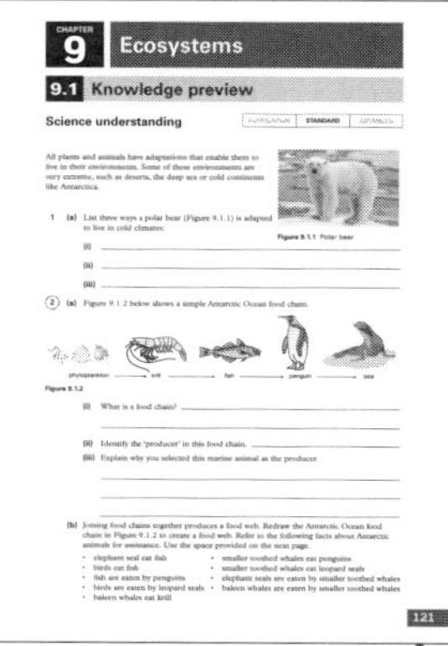
CHAPTER 9 Ecosystems

9.1 Knowledge preview

Science understanding

Be literate

Newly improved **literacy reviews**, in consultation with our Literacy Consultant Dr Trish Weekes, provide a deeper and broader range of language building tasks. Every chapter concludes with a literacy review which focuses on building a deeper understanding of key terms supporting students to correctly apply key terms from the topic.

Be set

Visit www.pearsonplaces.com.au to enjoy the benefits of the following digital assets and interactive resources to support your learning and teaching:

- New interactive activities and lessons
- New Untamed Science videos
- Web destinations
- Student investigation templates and teacher support
- New STEP-UP student book and activity book chapters with answers at Years 9 and 10
- Full answers to all Student Book and Activity Book questions
- SPARKlabs
- Risk assessments
- Full teaching programs and curriculum mapping audits
- Chapter tests with answers

Worksheets

Each worksheet is classified according to the degree with which it deals with curriculum understandings.

- **Foundation** indicates the focus is on the basics like terminology.
- **Standard** indicates a focus on the core ideas, understandings and skills.
- **Advanced** indicates transfer and extension of core science understanding and skills to new or more sophisticated situations.

Teachers may use this tool to **differentiate** the worksheets allocated to students. They may select worksheets for students based on whether basic, core or extension exercises are required. The categories do not indicate the degree of difficulty of tasks on the worksheet. A worksheet labelled advanced may have tasks ranging from lower level through to higher-level thinking.

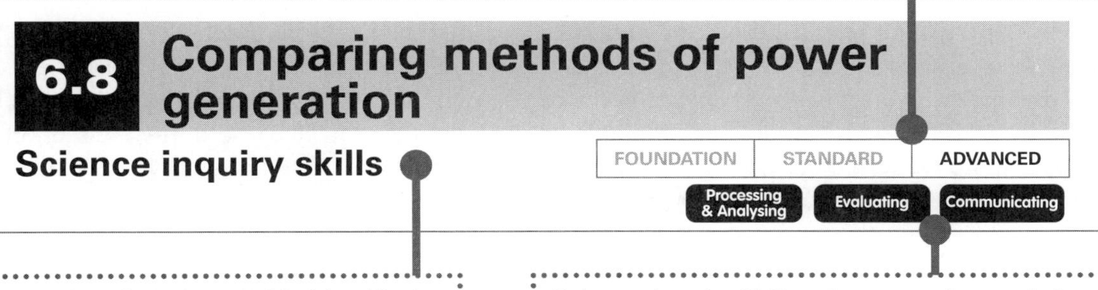

The main science **strand** is identified for each worksheet: Science Inquiry Skills, Science Understanding, Science as a Human Endeavour.

Science Inquiry Skills relevant to the worksheet are identified: Questioning and Predicting, Planning and Conducting, Processing and Analysing, Evaluating, Communicating.

Each question in a worksheet is identified according to the degree of difficulty. Bloom's taxonomy is used as the basis for question classification. The classification is intentionally subtle and unobtrusive. This tool may be used to differentiate between students, matching questions to their levels of ability.

Questions with a straight number indicate a remembering or understanding lower-order question.

Questions with a circle around the question number indicate an analysing or applying middle-order question.

Questions with a square around the question number indicate a creating or evaluating higher-order question.

1 Read through the words in the vocabular
(a) List the words that you know and wr

(5) Identify the solutes that did not dissolve in

[7] Discuss how this story illustrates that scie

An innovative tool for students to quickly and easily reflect on their understanding of each worksheet. The teacher may use the student responses as a formative assessment tool. At a glance, teachers may assess which topics and which students need intervention for improvement.

RATE MY UNDERSTANDING
Shade the face that shows your rating   

Science toolkit

Thinking scientifically

Designing an experiment

When designing an experiment you need to consider a number of factors. These include:

- making it a fair test
- eliminating errors and mistakes
- managing safety and risks
- working with data
- writing the report.

Making it a fair test

An experiment is a fair test if there is only one variable that changes (the independent variable) while all other relevant variables are kept constant and don't change (the controlled variables). Always assess which variables are relevant and which are not. For example, in an experiment investigating plant growth, the amount of water given to the plants is likely to have an effect on the results so should be kept constant—the same amount of water; the same frequency of watering. The colour of the pots that the plants are growing in is unlikely to be important.

Understanding variables

Variables may be dependent, independent or controlled. Figure 0.1 shows how these three types of variables affect an experiment.

Variables in a plant growth experiment

Hypothesis: If plants are grown in green light they will die.

Suitable experiment to test the hypothesis: grow some plants in green light and other plants in normal white light.

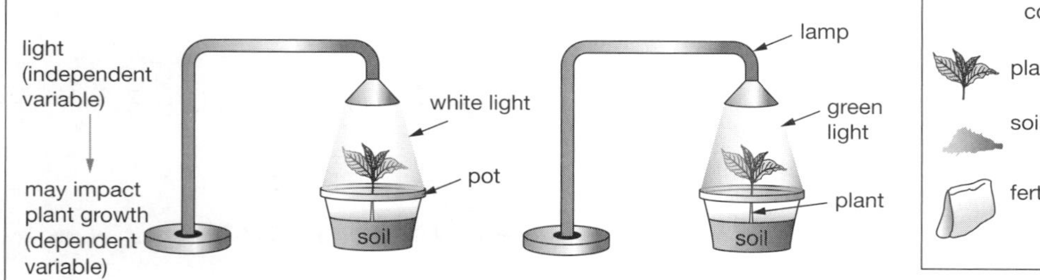

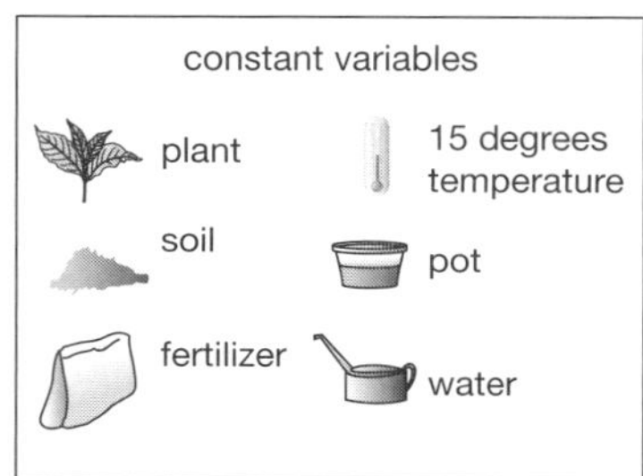

The **dependent variable** is the factor being tested and measured. It is what happens as a result of the effects of the independent variable. In this example the dependent variable is **how well the plants grow**.

The **independent variable** is the factor you choose to change or manipulate. The independent variable in this example is **the colour of the light**.

The **controlled variables** are all other factors which must be taken into account and are kept constant throughout the experiment. In this example they include **temperature**, **amount of fertiliser** applied, **size of pots**, **soil quality**, **size and type of plants** at start of experiment.

Figure 0.1

Eliminating errors and mistakes

Errors and mistakes may affect results. It is important to identify and, as far as possible, eliminate errors and mistakes, to get accurate results.

Mistakes occur when the experimenter does something incorrectly. Mistakes usually occur because the experimenter was not careful enough with the equipment, or because they failed to accurately follow the method. Mistakes are not scientific errors. Errors occur when taking measurements and may be systematic or random.

Common mistakes and how to overcome them

Common mistake	Ways to overcome mistake
incorrect method	• carefully read instructions • understand purpose of the experiment • carry out method steps in the correct order • do not take short cuts
inaccurate measurements	• measure amounts of chemicals carefully • do not spill chemicals when measuring them or results will be wrong • do not forget to add chemicals as instructed
wrong calculations	• make sure correct formula is used • check that there are no arithmetic mistakes • be careful with placement of decimal points in numbers

Common errors and how to overcome them

Common errors	Examples	Ways to overcome errors
Systematic errors may be due to a particular mistake being made every time or if an instrument is not calibrated correctly. As these types of errors are the same all the time they change the total results but they do not affect the general trends in the results. They do not generally cause inaccurate conclusions to be drawn. These errors are difficult to detect.	• a set of scales which always weighs 0.5 g too heavy • a thermometer which always reads 1°C less than the actual temperature	• Check instrument calibration before starting an experiment. • Set instrument back to zero if possible. • Ask teacher to check calibration if you cannot do it.
Random errors may be due to fluctuations in equipment and inconsistencies in interpretation of readings. These errors are easier to identify as two readings for the same measurement will greatly vary. These errors can have the most serious impact on the experimental results as they can change the trends which will lead to incorrect conclusions.	**Instrument increments** To measure 26 mL of water, a beaker and a measuring cylinder were used (Figure 0.2). The gradations on the measuring cylinder mean that the amount of water is 26 ± 0.5 mL. We can be sure that the measurement is between 25 and 26 mL. The measurement in beaker would be no more accurate than 25 mL ± 20 mL. 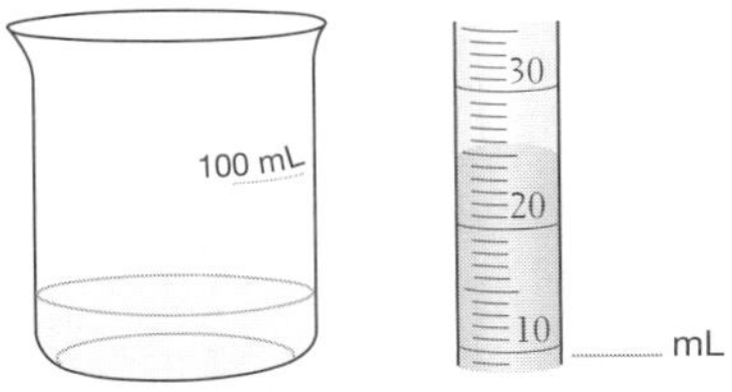 **Figure 0.2**	• Obtain more than one measurement to check results. • Repeat the experiment and calculate the mean of all the values. • When a measurement reading is made that is very different from other readings, repeat the reading and/or compare with other experimenters.

Two common random errors are related to instrument reading—increments on instruments and parallax.	**Parallax** This error (Figure 0.3) is caused by reading dials and instruments from an angle at the side, giving inaccurate readings. 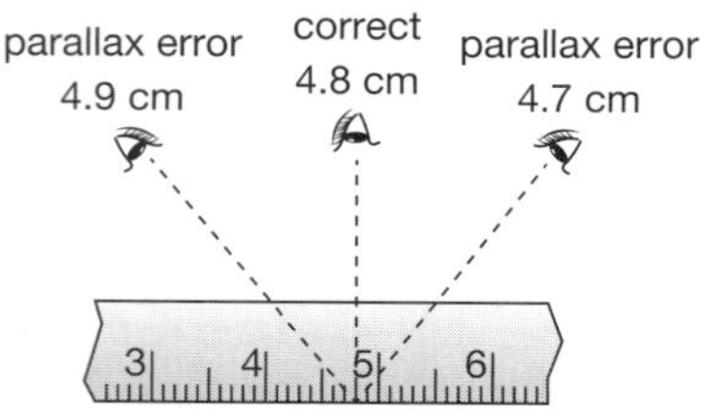 **Figure 0.3**	• When measuring liquids the piece of equipment with the smallest possible measuring increments should be used as this reduces the need for estimating. • Always read any measurement straight on at eye level and directly in front of the instrument at right angles to measurement.

Managing safety and risks

Before any experiment is undertaken, all possible sources of injury to people or harm to the environment must be identified so that these risks can be minimised or eliminated. There are five levels of safety as shown in Figure 0.4. The safety levels are ranked in order of importance. The school and teacher are responsible for reducing the most important risks. You, the student scientist, can take measures to reduce the risks at the bottom of the hierarchy.

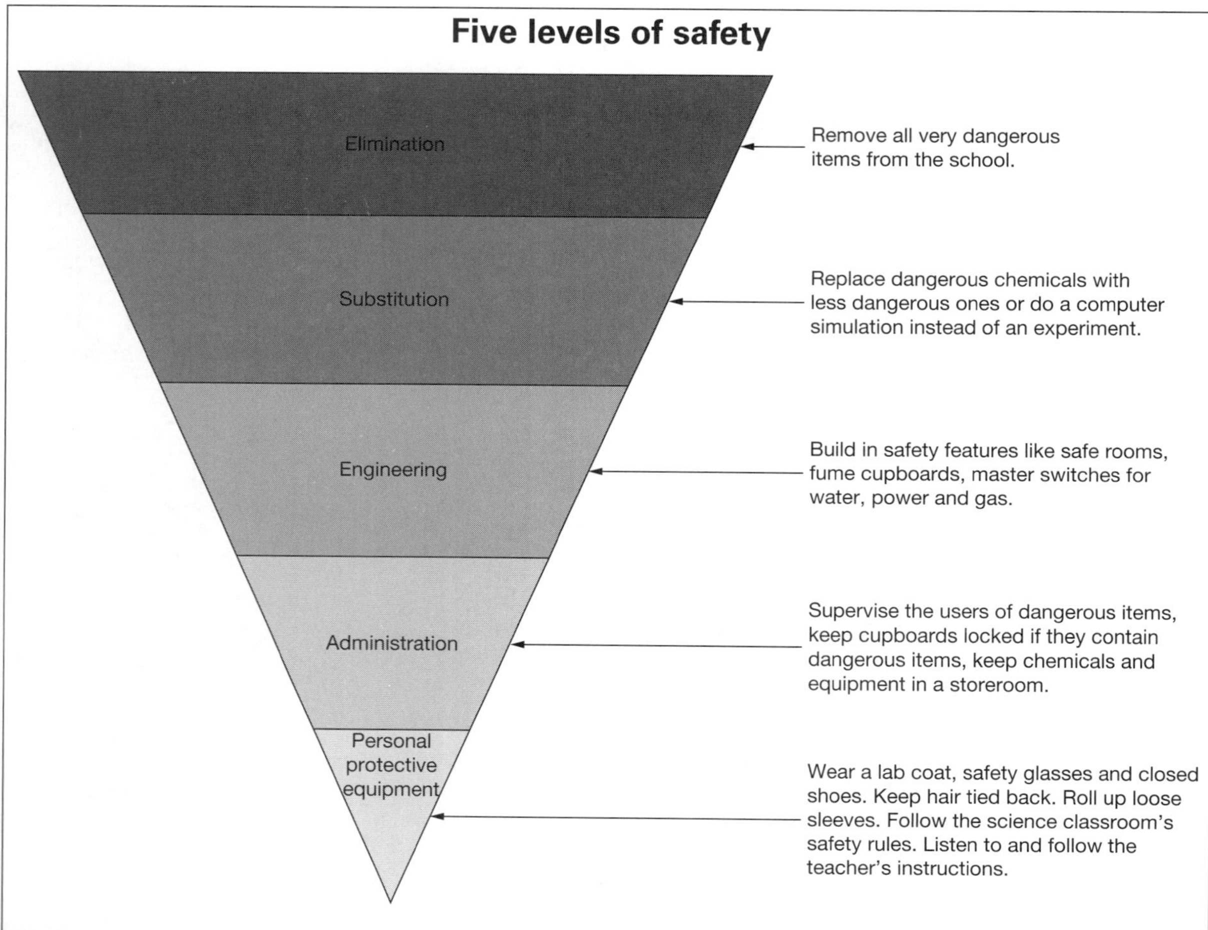

Figure 0.4

Checking for potential risks

In order to check for potential risks, a risk assessment needs to be completed. A sample Risk Assessment Form is shown in Table 0.1.

Risk Assessment Form	
What activity will you be doing?	
Title or description of the investigation	
What are the risks involved?	**How can you REDUCE the risks?**
List the equipment you will be using.	State how you will use each piece of equipment.
List the chemicals you will be using.	State how you will carefully use each chemical. State how you will carefully dispose of the chemicals.
List any other possible risks.	State how you will reduce each of these risks.

Table 0.1

Working with data

The data collected when conducting an experiment can be **qualitative** or **quantitative**. Graphs are used as a visual way of presenting data. The type of graph that will be required depends of the type of data that has been collected.

	Qualitative data	Quantitative data
Description	• qualitative data is a description • an example is 'silk is smoother than wool' • is given either as a short paragraph or may be presented in a table • falls into two categories: continuous and discrete (also called discontinuous) • discrete data can be graphed using bar or column graphs or pie charts • continuous data is generally graphed using a line graph	• data using numbers • an example is the heights of the plants in centimetres

Drawing graphs using Excel

Widespread access to computers means that most practical reports will be produced using computer programs such as Word, Google Docs and Excel. Unfortunately, many students are unfamiliar with the details of these applications, especially Excel, and so they submit work which does not conform to the required graphing conventions.

The tomato plants, in the experiment 'An investigation into the effect of the fertilisers 'Growmore' and 'Happy Plants' on the growth of tomato plants' on pages xv–xvii, provide continuous data. The heights of the plants were only recorded every second day, but we know the plants had a height and were continuing to grow between the times the measurements were taken. With the right equipment we could have recorded the heights continuously throughout the 20 days of the experiment.

Creating a line graph for continuous data

Step 1: To draw a line graph in Excel, open a new spreadsheet and enter your data into the cells. Key each heading into individual cells along a single row. In the example shown in Figure 0.5, the headings are at the top of each of the data columns. Key in your data (numbers) under the correct headings in each column. Each number is keyed into its own cell. Make sure you save your spreadsheet regularly.

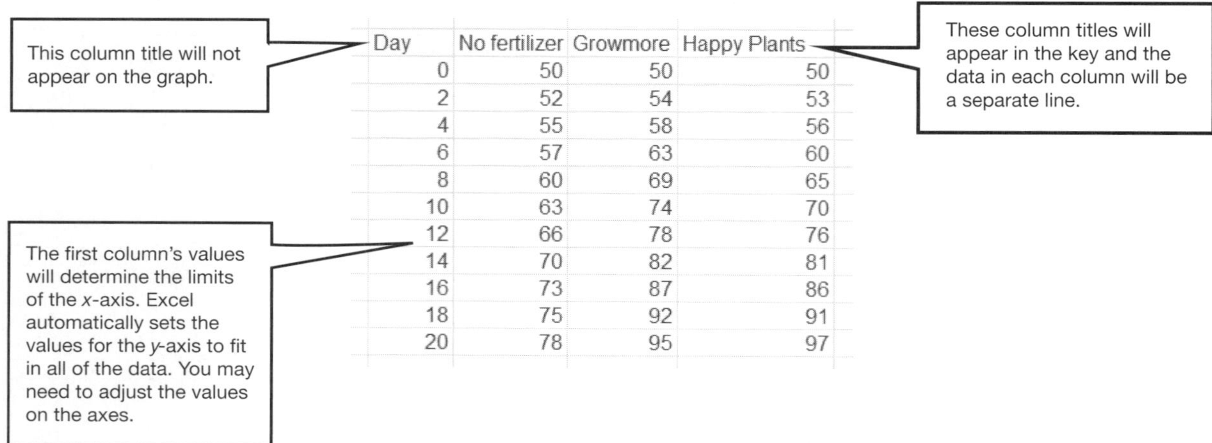

Day	No fertilizer	Growmore	Happy Plants
0	50	50	50
2	52	54	53
4	55	58	56
6	57	63	60
8	60	69	65
10	63	74	70
12	66	78	76
14	70	82	81
16	73	87	86
18	75	92	91
20	78	95	97

Figure 0.5

Step 2: As shown in Figure 0.6, highlight the data you want included in your graph. Move the mouse pointer to the top menu and left click on 'Insert' to select it. Next, move the mouse pointer to the charts menu to the right and select 'Scatter'. From the drop down 'Scatter' menu, select 'Scatter with smooth lines'. Your graph will then appear.

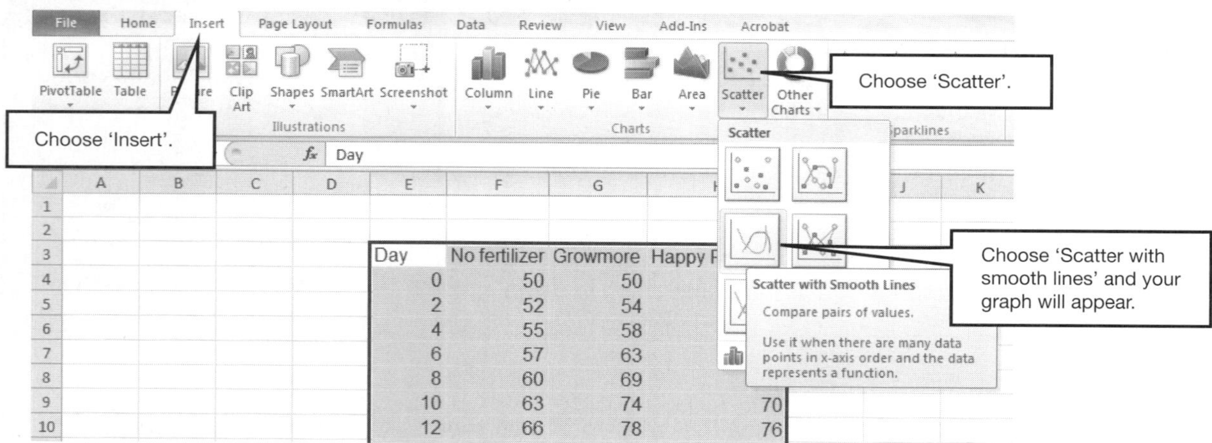

Figure 0.6

Step 3: To change the values on either of the axes, click on one of the numbers to highlight the axis; then, keeping your cursor on the number, right click and choose 'format axis' (see Figure 0.7).

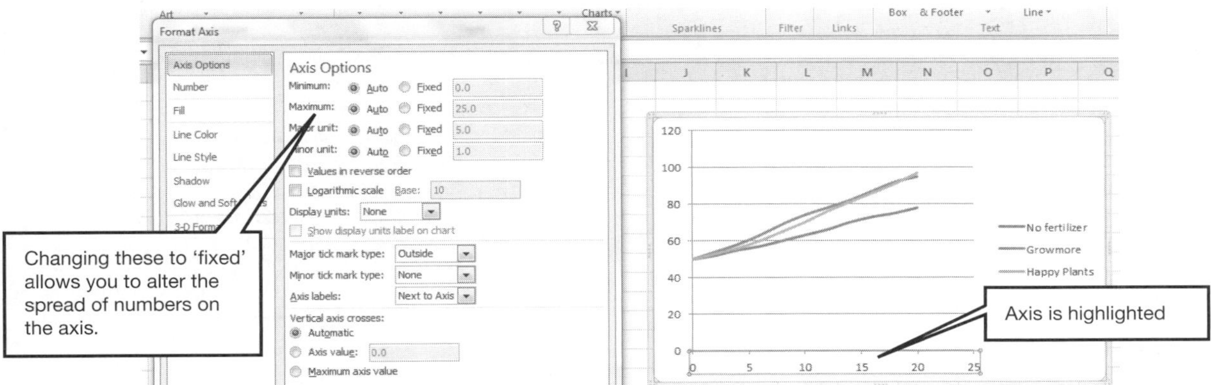

Figure 0.7

Step 4: Finally, label your axes and give your graph a title as shown in Figure 0.8.

Click anywhere inside the border of your graph to bring up the 'Chart Tools' menu which will appear at the top of the screen.

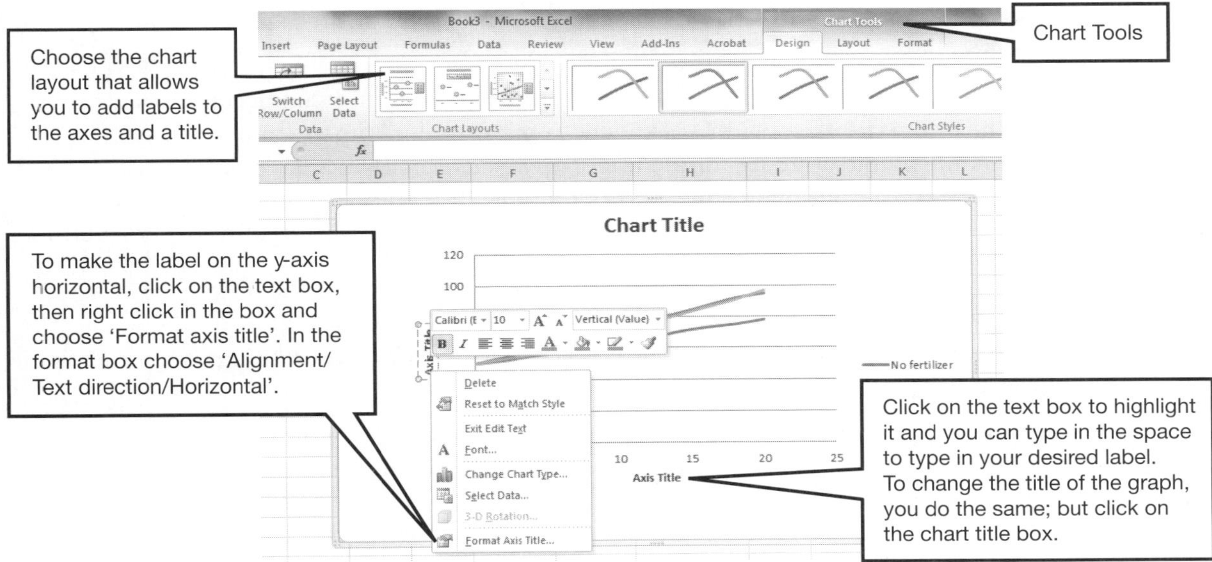

Figure 0.8

Creating a column graph for discrete (discontinuous) data

Discrete (discontinuous) data is usually plotted on a column graph. Data is discrete if there was no data between the readings that were taken. If more than one set of data is to be plotted, they can be done in two blocks or the columns can be drawn side by side.

Highlight the data you want to include in the graph. Click on 'Column' from the 'Charts' menu, and your graph will appear. Follow steps 3 and 4 used for 'creating a line graph …' to format the axes and label the column graph. Figure 0.9 shows how you can change the colour of the columns.

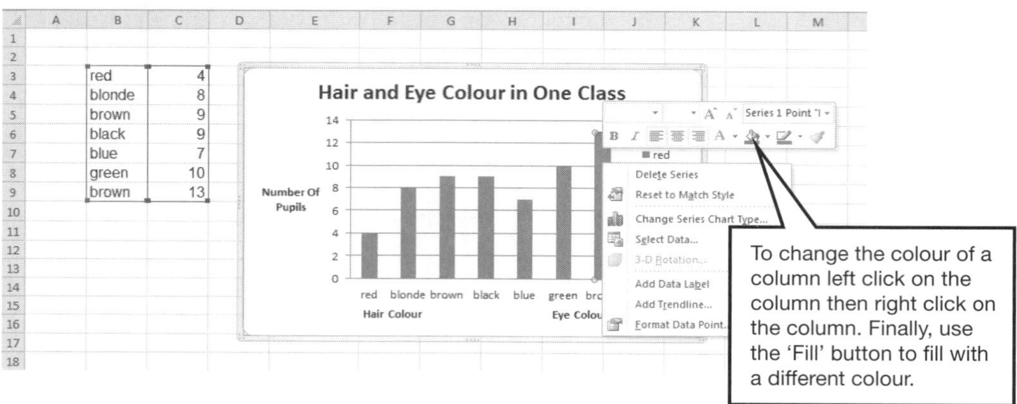

Figure 0.9

Figure 0.9 shows eye colour and hair colour of students in one particular class on the same set of axes so that they can be compared.

Writing the report

Scientists communicate their findings and ideas about the significance of their experiment's results in a practical report which is published in a scientific journal. Since practical reports are designed for publication, they are formal pieces of writing following rules associated with their format and content. Practical reports should not contain informal language and should be written in an impersonal style.

At a school level, some of those rules are relaxed and others are not enforced because you are learning not only how to write a report but also how to perform experiments. For example, students do not include an introduction of the underlying scientific theory unless specifically asked to do so. A published report presents results as either a table or a graph. Students include both a table and graph of results so teachers can make sure the students can accurately convert tables into graphs.

A practical report model

The format of a practical report is fairly standard. It consists of a series of headings with specific information required under each heading.

A student designed and performed an experiment to determine the effect of two different fertilisers on the growth of tomato plants. The student then wrote a practical report which is shown on the pages that follow (Figure 0.10).

Title: The title is not a catchy slogan. It should give a clear description about what experiment was performed.

Introduction: The introduction should set the scene for the experiment. It should briefly discuss the reasons for doing the experiment and the scientific theory which underlies the experiment.

Hypothesis: A hypothesis is a prediction about what will happen. It should state what you think will happen.

Safety: Every experiment has some risks associated with it. A risk assessment should be performed and instructions as to what those risks are, and how they can be minimised, should be given. Figure 0.5 on page xi provides a detailed risk assessment sheet to help you with this task.

This group is the control. It is the group we will compare against to determine whether the fertiliser made any difference.

Aim: The aim should clearly state what you hope to find out by doing the experiment.

Materials: All materials needed to perform the experiment should be listed. It is not necessary to include laboratory coats and safety glasses; although they should be worn when any experiment is performed.

An investigation into the effect of the fertilisers 'Growmore' and 'Happy Plants' on the growth of tomato plants.

Introduction

A focus on healthy eating and fresh foods has resulted in an increase in the number of people growing their own fruits and vegetables. All plants need minerals and other nutrients to grow to their optimum in size.
Most gardeners use fertilisers on their plants in order to make sure the plants have the nutrients that they need. Fertilisers can be expensive to buy so it is important to determine if they are actually increasing the growth of the plants.

Aim

To determine whether using fertilisers 'Growmore' and 'Happy Plants' increases the growth of tomato plants.

Hypothesis

If fertiliser is used, then the tomato plants will grow faster.

Safety Considerations

Some fertilisers and some soils can be infected with ***Legionella*** bacteria so care should be taken not to breathe in or ingest soil or fertilisers. Gloves should be worn and hands washed thoroughly after plants and/or fertilisers are handled.

Materials

- 90 identical tomato seedlings (50 mm tall)
- 90 identical pots
- 1 bag of soil sufficient to fill all 90 pots
- 12 bottles of Growmore liquid fertiliser
- 6 boxes of Happy Plants dry fertiliser
- 3 soil dampness probes
- ruler
- gloves
- scales

Method

1 Pot the tomato plants into the 90 pots, with one plant per pot, using the soil. Ensure each pot contains the same amount of soil by weighing out the same amount for each plant.
2 Divide the pots into three groups of 30 plants.
3 Label the first group 'water only'. Only give this group water as needed.
4 Label the second group 'Growmore' and the third group 'Happy Plants'.

5 Groups 2 and 3 should be fertilised with their respective fertilisers. Follow the directions on the packets carefully.

6 Water all plants when the moisture probe indicates that this is needed.

7 All plants should be exposed to the same amount of light and all other conditions, except the fertiliser, should be kept the same. Throughout the experiment the moisture content of the soil should be monitored in order to keep this constant.

8 Using the ruler, measure and record the heights of the plants (in mm) in each group at the start of the experiment. Then measure and record the heights (in mm) of the plants in each group every second day for 20 days.

These groups are the experimental groups. The only difference that there should be between these and the control is the fertiliser.

Results

Results: The results section should clearly show the observations that you have made. Scientists generally present their data as either a table **or** a graph, but often your teacher will ask you to do both so that they can check how well you can convert your raw data into a graph.

Table 1: Mean growth over 20 days, in millimetres, of tomato plants with and without fertiliser.

Tables should have a table number and a title which describes how the data in the table was generated.

Days of growth	Mean Height of Plants with each of the Treatments (mm)		
	No fertiliser	Growmore fertiliser	Happy Plants fertiliser
0	50	50	50
2	52	54	53
4	55	58	56
6	57	63	60
8	60	69	65
10	63	74	70
12	66	78	76
14	70	82	81
16	73	87	86
18	75	92	91
20	78	95	97

All of the columns in a table must have a title. Units should be included in the table.

The results shown in the table are the mean heights for each group on each day. For each group the heights of the 30 plants were measured and then the mean was calculated for each day of the experiment. You should have a logbook in which you collect and keep all of your raw data. Your logbook should also include all of your calculations.

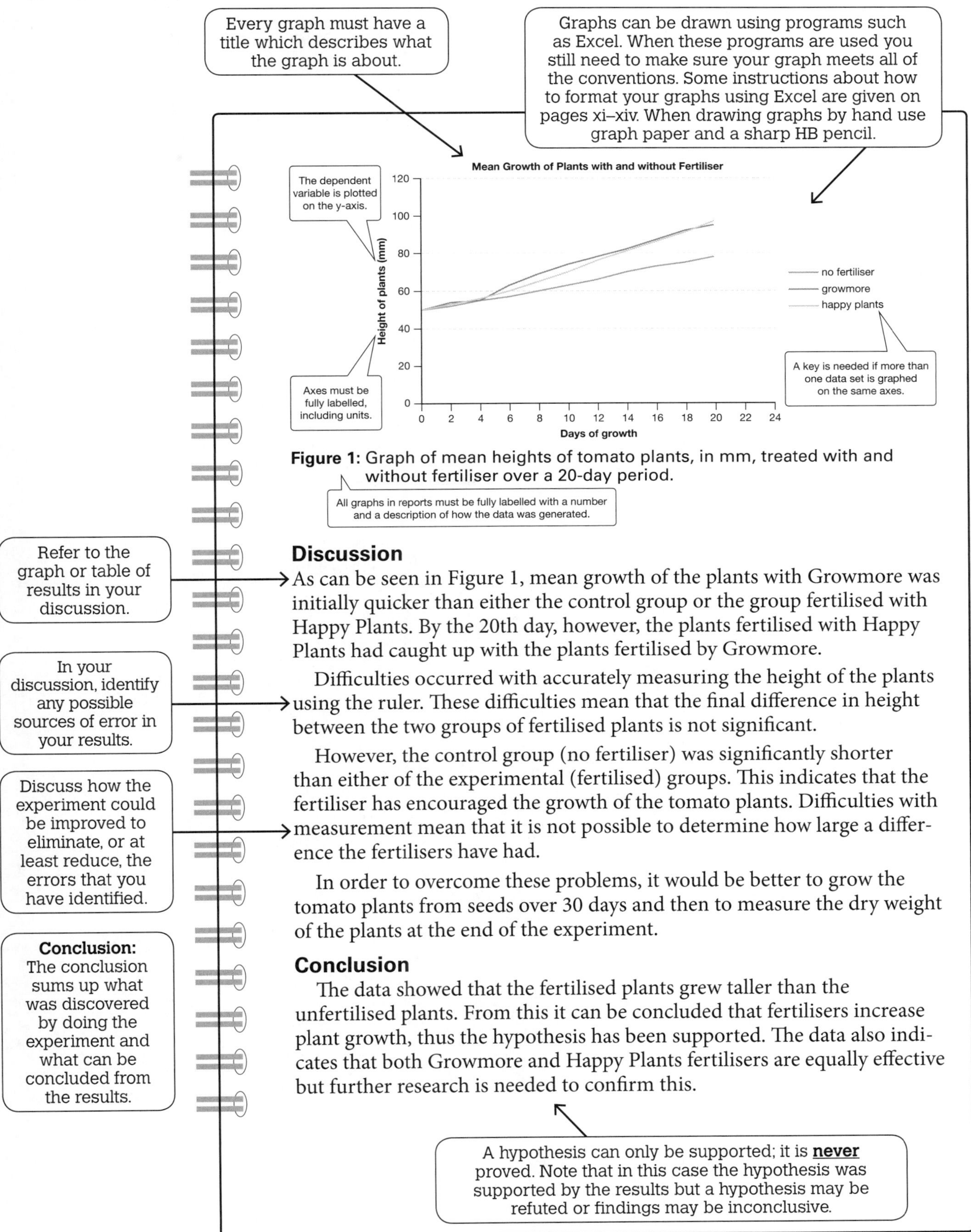

Figure 1: Graph of mean heights of tomato plants, in mm, treated with and without fertiliser over a 20-day period.

Discussion

As can be seen in Figure 1, mean growth of the plants with Growmore was initially quicker than either the control group or the group fertilised with Happy Plants. By the 20th day, however, the plants fertilised with Happy Plants had caught up with the plants fertilised by Growmore.

Difficulties occurred with accurately measuring the height of the plants using the ruler. These difficulties mean that the final difference in height between the two groups of fertilised plants is not significant.

However, the control group (no fertiliser) was significantly shorter than either of the experimental (fertilised) groups. This indicates that the fertiliser has encouraged the growth of the tomato plants. Difficulties with measurement mean that it is not possible to determine how large a difference the fertilisers have had.

In order to overcome these problems, it would be better to grow the tomato plants from seeds over 30 days and then to measure the dry weight of the plants at the end of the experiment.

Conclusion

The data showed that the fertilised plants grew taller than the unfertilised plants. From this it can be concluded that fertilisers increase plant growth, thus the hypothesis has been supported. The data also indicates that both Growmore and Happy Plants fertilisers are equally effective but further research is needed to confirm this.

Figure 0.10 A model of a practical report

CHAPTER 1

Science skills

1.1 Knowledge preview

Science understanding

FOUNDATION | STANDARD | ADVANCED

The scientific method starts when you question something and then investigate why it happens by conducting a scientific investigation. There is a set process to follow in scientific investigations. Complete the following questions to assess your understanding of this process.

The hypothesis

1 What is a hypothesis?

____

____

(2) Why is it important to write a hypothesis?

____

____

____

(3) Explain the difference between an independent and a dependent variable.

____

____

____

Variables

4 What is a variable?

____

____

5 Identify the independent and dependent variables in the following hypotheses. Draw ovals around the independent variables and rectangles around the dependent variables.

(a) If a cup of hot chocolate has a lid on it, then it will stay hot for a longer period of time.

(b) Because thin candles have less wax to burn, they will burn faster than thick candles.

Procedure

6 Read the purpose and procedure for the experiment that follows:

An investigation into the effect of pH on seedling growth experiment

Purpose To investigate the effect of pH on seedling growth.

Hypothesis ______________________________

Procedure

1. Germinate 20 seed pods on damp cotton wool.
2. Choose 12 seedlings with a height of about 12 mm.
3. Plant each seedling in a pot of the same size. For each pot, use 80 g of quality potting mix and water with 12 mL of tap water.
4. Label each pot with the pH treatment the soil will receive: 4 pots with pH 5, 4 pots with pH 7 and 4 pots with pH 9.
5. Weigh each pot to the nearest 0.1 g. Draw up a data table and record the results for each pot in the column for day 0.
6. Keep plants in the same position where light is available to maintain lighting conditions.
7. Reweigh the seedlings in their pots 2 days later. Record the results for each pot in the column for day 2.
8. Immediately after each weighing, give each plant 10mL of water at the appropriate pH according to the label on the pot.
9. Repeat steps 5 and 6 every 2 days for the next 10 days.
10. Repeat steps 1 to 8 twice to reduce the chance of variability between trials.

pH 5 6/4/2016 (×4)
pH 7 6/4/2016 (×4)
pH 9 6/4/2016 (×4)

(a) State the independent variable for the experiment.

____

____

(b) State the dependent variable for the experiment.

____

____

(c) Write a hypothesis for the experiment.

____

____

(d) List the controlled variables stated in the procedure.

____

____

____

(e) Explain the importance of controlling all variables except the dependent variable.

____

____

____

1.2 Variables and hypotheses

Science understanding

FOUNDATION | **STANDARD** | ADVANCED

1 Define the following terms:

(a) Research question ______

(b) Hypothesis ______

(c) Independent variable ______

(d) Dependent variable ______

(e) Controlled variable ______

2 Read through the following research questions and identify the independent and dependent variable for each:

(a) Does the mass of an object affect the rate at which it falls?

(i) independent variable ______

(ii) dependent variable ______

(b) How does light affect the growth of a plant?

(i) independent variable ______

(ii) dependent variable ______

1.2 Variables and hypotheses

3 In science it is important to be able to write a hypothesis that can be tested. Write a hypothesis for the following research questions:

(a) How does exercise affect heart rate?

(b) Does salinity affect the germination of seeds?

4 Without controlling variables, an experiment will not be a fair test. Complete the following table to demonstrate your understanding of controlled variables in an experiment.

Research question: Which glue stick is the most effective in gluing paper to cardboard?

Variable being controlled	How it will be controlled?	How it will affect the experiment if it is not controlled?

RATE MY UNDERSTANDING
Shade the face that shows your rating

1.3 Observing and making predictions

Science inquiry skills

FOUNDATION | **STANDARD** | ADVANCED

Questioning & Predicting

1 List five observations of the candle pictured in Figure 1.3.1:

(a) ______________________________

(b) ______________________________

(c) ______________________________

(d) ______________________________

(e) ______________________________

Figure 1.3.1

2 At the front of the class, a teacher lights and burns a candle for 10 minutes, then extinguishes the flame using a candle snuffer as shown in Figure 1.3.2. White smoke is observed coming from the top of the candle after it is extinguished.

The students are asked:

"When a candle is lit is it the candle wax that burns, the wick that burns or is it both the wax and the wick that burn?"

(a) Write a hypothesis for the question you believe to be correct.

____

____

Figure 1.3.2

(b) A hypothesis is written based on research or experience. What makes you think the option you have chosen is correct?

____

____

(c) The teacher relights the candle, puts it out, and then holds a match in the white wax vapour (smoke) that is coming off the candle. The wax vapour catches light and the flame travels down the smoke and relights the candle. Does this extra information change your prediction? Justify your answer.

____

____

____

____

____

candle snuffer (*n*) a long-handled, metal tool with a cone-shaped head used to put out a candle flame

extinguish (*v*) to put out a flame or fire

inference (*n*) a conclusion reached based on reasoning and evidence

sublime (*v*) to change a solid substance into a vapour by heating

vapour (*n*) moisture or a substance suspended in the air and visible as smoke

1.3 Observing and making predictions

3 A student takes a beaker of carbon dioxide gas and pours it over the top of the lit candle. The flame flickers and goes out.

Which of these statements are inferences supported by the student's observations?

(a) carbon dioxide is heavier than air

(b) carbon dioxide is a white gas

(c) carbon dioxide extinguished the flame

(d) carbon dioxide flowed out of the beaker and over the candle

(e) carbon dioxide is soluble in water

(f) carbon dioxide has a strong smell

____

____

4 A teacher has a bowl of dry ice subliming on the front desk. She lights three candles and puts them on the desk around the bowl. After a few minutes all three candles have gone out.

(a) Predict what happened to the candles.

____

____

(b) Explain how you test your prediction.

____

____

RATE MY UNDERSTANDING
Shade the face that shows your rating

1.4 Designing a controlled experiment

Science inquiry skills

FOUNDATION	STANDARD	ADVANCED

Planning & Conducting

The coaches of the national swimming squad want to test a new swimsuit. To do this they will compare a swimmer wearing the usual standard swimsuit competing against a swimmer wearing the new swimsuit.

1 Design a controlled investigation to test the effectiveness of different swimsuits in the pool. Describe the research question, hypothesis, variables (controlled, independent and dependent), materials, risk analysis, procedure and results below.

Figure 1.4.1 There are a variety of different swimsuits worn by swimmers. Polyurethane-enhanced swimsuits allow faster and smoother movement through the water.

Research question

____

____

____

____

____

Hypothesis ______________________________

____

____

____

____

Variables: independent ______________________________

____

____

Variables: dependent ______________________________

____

____

Variables: controlled ______________________________

____

____

____

____

1.4 Designing a controlled experiment

Materials ______

Risk analysis ______

Procedure ______

Results Draw a results table that could be used to collect results for this experiment.

RATE MY UNDERSTANDING
Shade the face that shows your rating

1.5 Processing and evaluating data

Science inquiry skills

FOUNDATION	**STANDARD**	ADVANCED

Processing & Analysing | Evaluating

Sophia and Claude did an experiment with three trials. The purpose was to find out the best temperature for tomato seeds to germinate. Results from their experiment workbook are shown here.

Number of seeds that germinated out of 50			
Temperature °C	Trial 1	Trial 2	Trial 3
0	0	0	0
5	0	0	0
10	40	42	20
15	14	16	15
20	8	8	8
25	6	5	6
30	6	5	6
35	9	8	10
40	0	0	0

1. Rewrite the table of data shown in Sophia and Claude's workbook in the space provided to the right, making improvements to the way data is organised and to the labels.

2. Construct a graph to represent the data, including a heading and fully labelled axes.

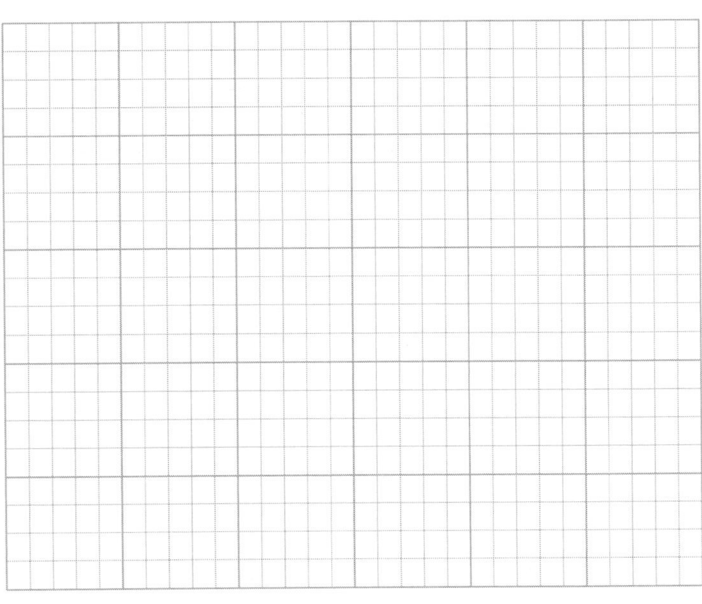

1.5 Processing and evaluating data

(3) State the independent variable for this experiment.

(4) State the dependent variable for this experiment.

(5) Identify three variables that should have been controlled in this experiment.

(6) Describe the trend shown by the data.

(7) Was sufficient data collected for this experiment? Explain your answer.

[8] Outline another experiment that could be done to inquire further into this investigation.

RATE MY UNDERSTANDING
Shade the face that shows your rating

1.6 Experiment report

Science inquiry skills

FOUNDATION	STANDARD	ADVANCED

Communicating

Writing up your experiment and the observations and data you collect may allow other scientists or students to build on your work. An experiment is usually written up under specific headings.

1. Using the flow chart below, summarise the steps in a scientific report and write a short explanation for what is required in each step. The first one has been completed for you.

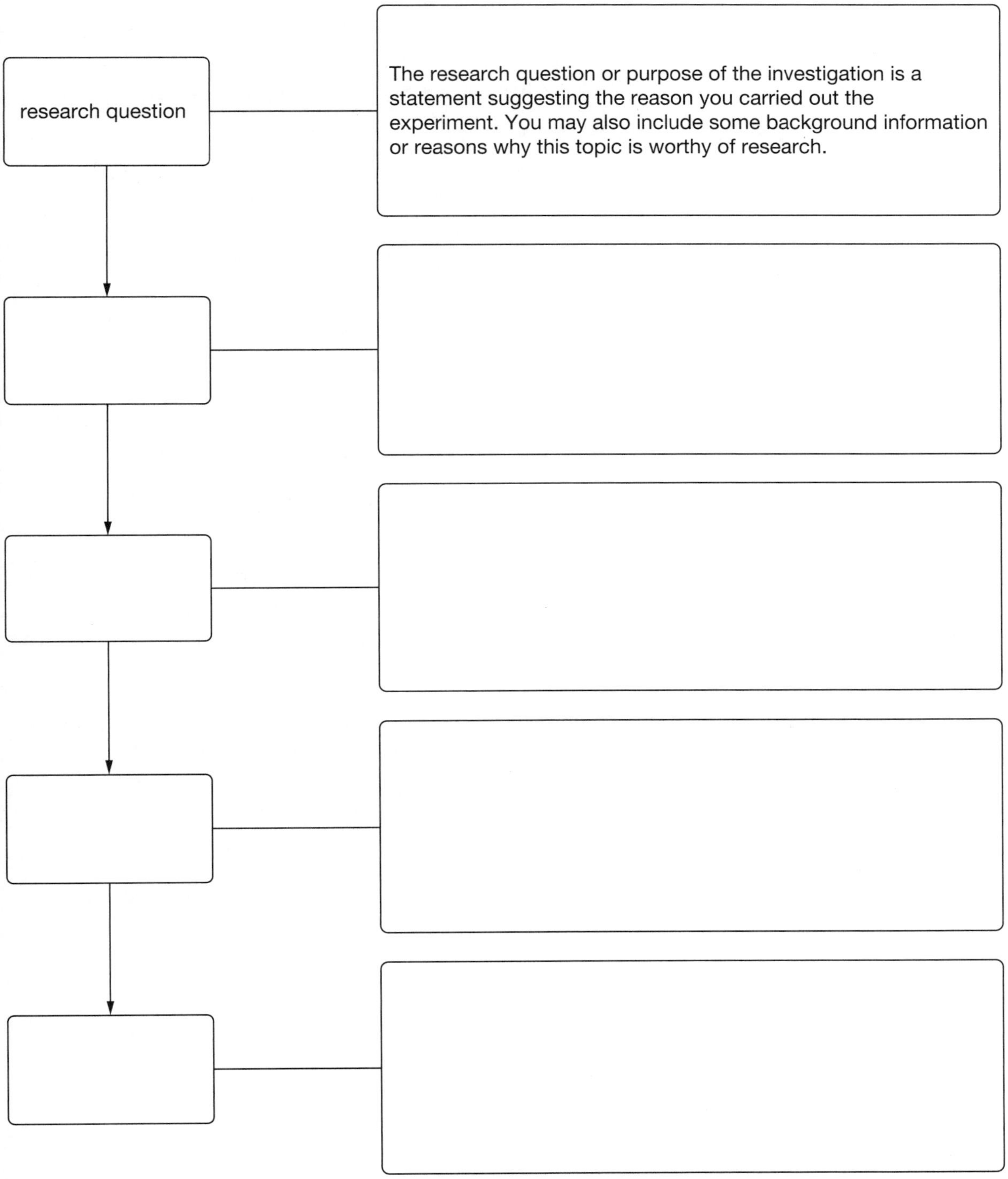

Step	Explanation
research question	The research question or purpose of the investigation is a statement suggesting the reason you carried out the experiment. You may also include some background information or reasons why this topic is worthy of research.

1.7 Literacy review

Science understanding

FOUNDATION	STANDARD	ADVANCED

1 Match the correct term with its definition by writing the term in the middle column next to its definition.

Incorrect term	Correct term	Definition
dependent variable		Something you notice or are aware of using your senses of sight, hearing, smell, touch and taste
procedure		A statement that is tested by doing an experiment
hypothesis		Consideration of the dangers of an experiment with the aim of reducing these by changing procedures or wearing protective clothing
conclusion		A variable that changes in response to changes in the independent variable and is measured in an experiment
observation		A factor that can affect the results of an investigation. This factor can be changed, kept the same or measured.
independent variable		Information that shows that something has occurred
risk assessment		A clear statement describing whether or not the results of the investigation support the hypothesis.
variable		This variable is kept the same during the investigation.
experiment		Results, often in the form of numbers
evidence		The variable that is changed in an investigation to see the effect it has on another variable
data		A test or procedure that is done to investigate a hypothesis
controlled variable		The method followed to carry out the investigation

RATE MY UNDERSTANDING
Shade the face that shows your rating

1.8 Thinking about my learning

Think back over your learning. Finish the sentences in the pentagon.

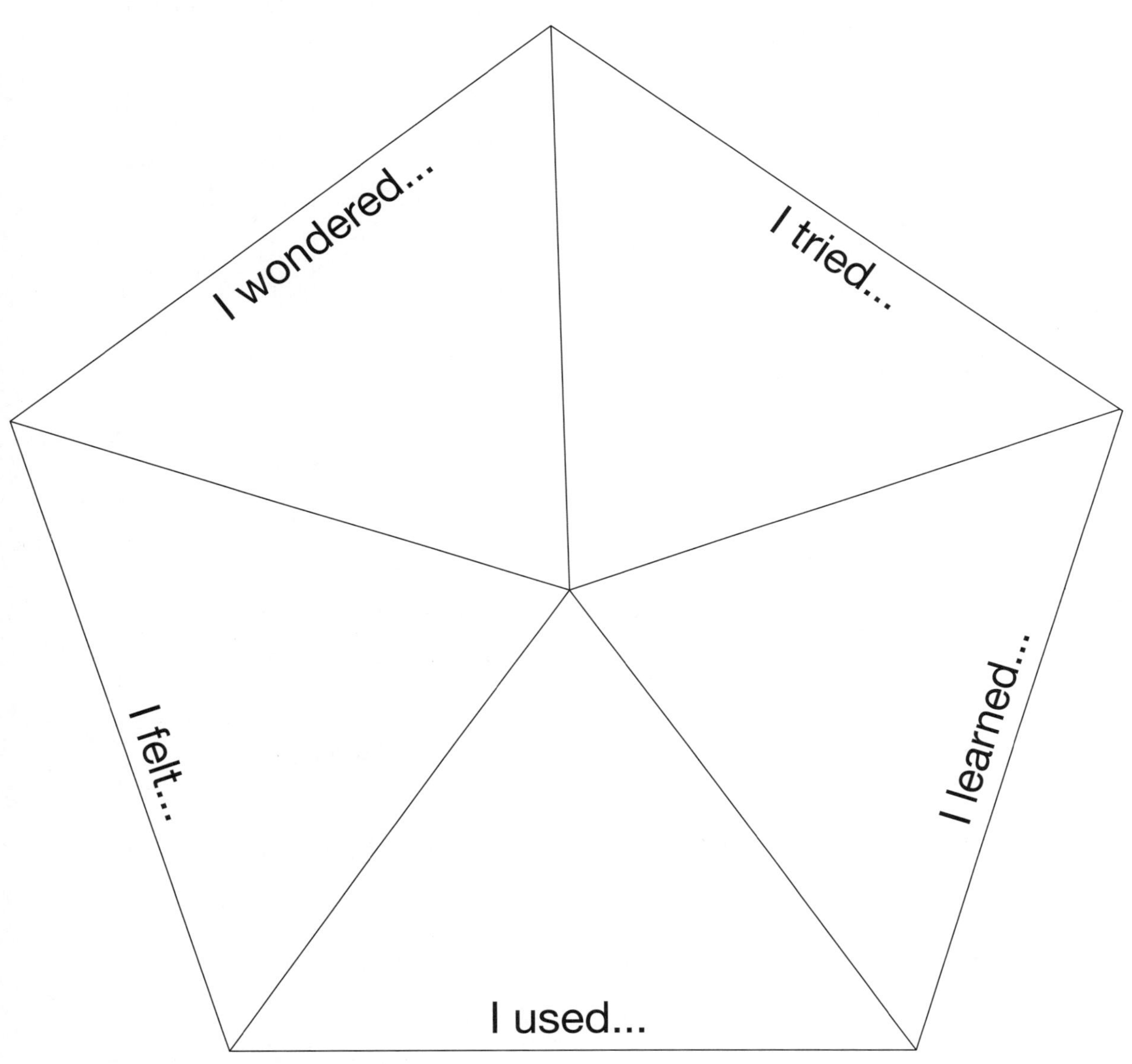

CHAPTER 2

Materials

2.1 Knowledge preview

Science understanding

FOUNDATION | STANDARD | ADVANCED

1 The diagrams below show an element, a compound and a mixture. Identify which is which by adding a label above each diagram.

(a) ______________

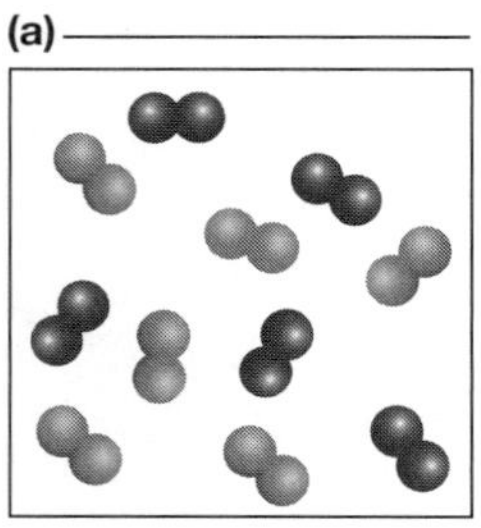

(b) ______________

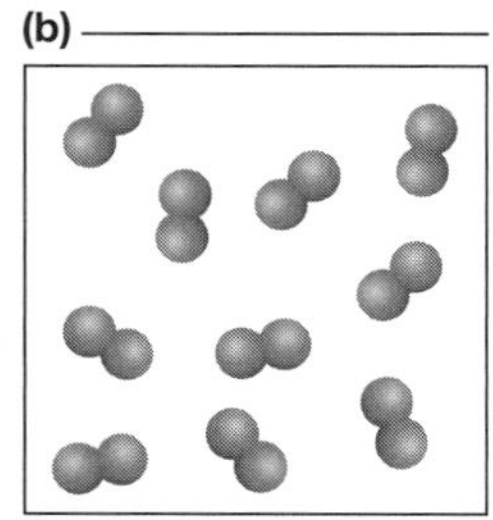

(c) ______________

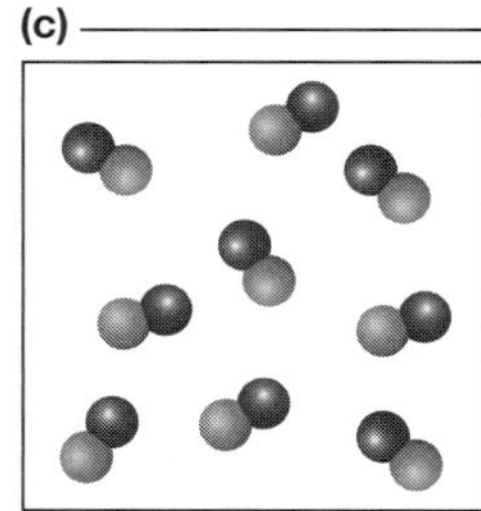

2 Define each of the following terms:

(a) element: ______________

(b) compound: ______________

(c) mixture: ______________

3 Use your existing knowledge about acids and bases to answer the following questions:

(a) What happens to red litmus paper when it comes in contact with a base?

____

(b) What colour would blue litmus paper turn when dipped in vinegar?

____

(c) An alkali is always a base but a base is not always an alkali. Suggest a reason why.

____

4 In this module you will be learning about metals. State three things you already know about metals:

(a) ______________

(b) ______________

(c) ______________

2.2 Atomic symbols

Science understanding

FOUNDATION | STANDARD | ADVANCED

Scientists use atomic symbols to communicate information about an atom. The atomic symbol is made up of the chemical symbol for the atom, the atomic number and the mass number. The atomic symbol of the carbon atom is shown in Figure 2.2.1

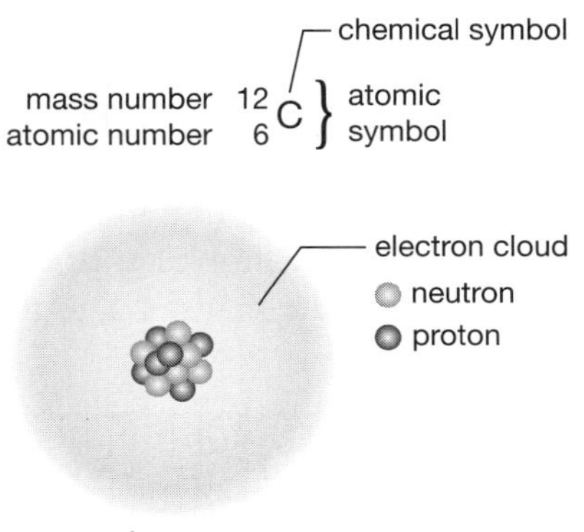

Figure 2.2.1

1 The atomic number is equal to the number of protons. The mass number is equal to the number of protons plus the number of neutrons. Count the number of protons and neutrons in the atoms below, then state the atomic number and mass number for each atom.

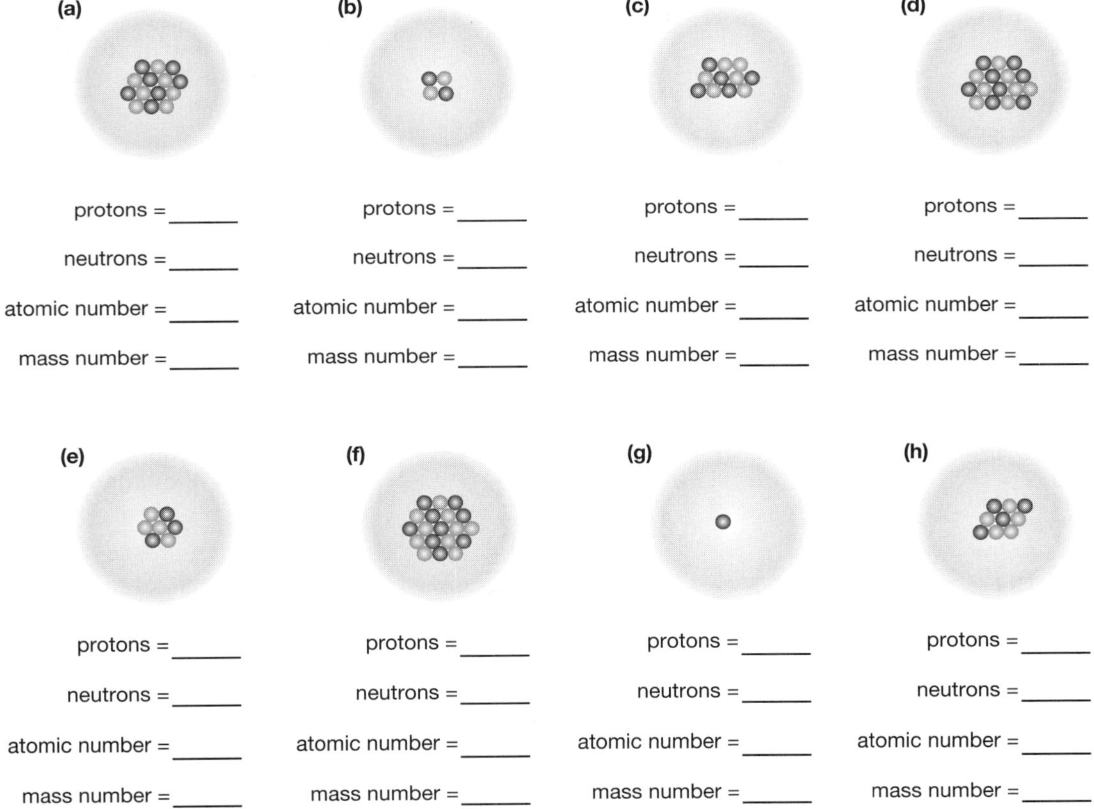

(a) protons = ______ neutrons = ______ atomic number = ______ mass number = ______

(b) protons = ______ neutrons = ______ atomic number = ______ mass number = ______

(c) protons = ______ neutrons = ______ atomic number = ______ mass number = ______

(d) protons = ______ neutrons = ______ atomic number = ______ mass number = ______

(e) protons = ______ neutrons = ______ atomic number = ______ mass number = ______

(f) protons = ______ neutrons = ______ atomic number = ______ mass number = ______

(g) protons = ______ neutrons = ______ atomic number = ______ mass number = ______

(h) protons = ______ neutrons = ______ atomic number = ______ mass number = ______

2 Match the letter for each atom in question 1 to its correct atomic symbol below.

$^{16}_{8}O$	$^{1}_{1}H$	$^{14}_{7}N$	$^{7}_{3}Li$
______	______	______	______
$^{4}_{2}He$	$^{9}_{4}Be$	$^{11}_{5}B$	$^{19}_{9}F$
______	______	______	______

RATE MY UNDERSTANDING
Shade the face that shows your rating

2.3 Atomic models

Science understanding

FOUNDATION	STANDARD	ADVANCED

Scientists, understanding of chemistry and of atoms has grown significantly over the last 2000 years and has led to the development of more detailed and accurate models of the structure of atoms.

1. The table below includes information about some of the models of atoms developed over time. Look at the jumbled columns of diagrams, names and descriptions of the different atomic models. Connect the diagrams, to their correct names and descriptions by drawing lines between them.

Atomic models

Diagram	Name of model	Description
	nuclear model	Atoms are made up of a positive nucleus that has both protons and neutrons. The nucleus is surrounded by electrons orbiting it like planets orbiting the Sun.
	solid-ball model	Atoms are made up of a positive nucleus that has both protons and neutrons. The electrons form shells around the nucleus.
	plum pudding model	Atoms are made up of a solid, positively charged nucleus surrounded by an electron cloud.
	electron cloud model	Atoms are made up of a positively charged ball with negatively charged electrons stuck to it.
	planetary model with neutrons	Substances are made of hard, ball-like building blocks called atoms that cannot be broken apart.

2. List the atomic models from the oldest to the most recent.

____

____

RATE MY UNDERSTANDING
Shade the face that shows your rating

2.4 Comparing alloys

Science inquiry skills

FOUNDATION | **STANDARD** | ADVANCED

Processing & Analysing | Communicating

Copper (Cu) and zinc (Zn) can be mixed together to form alloys of different strengths. For example, different strengths of brass can be obtained by mixing different proportions of copper and zinc. Table 2.4.1 shows the force that different alloys of copper and zinc can take before breaking.

Table 2.4.1

Brass with different proportions of copper and force needed to break it											
% Cu	0	10	20	30	40	50	60	70	80	90	100
Force ($\times 10^6$ N)	19	16	12	8	5	32	58	40	23	21	33

1. On the grid provided, construct a line graph of force (*y*-axis) plotted against the percentage of copper (*x*-axis). Connect the points together with a curve drawn smoothly through them.

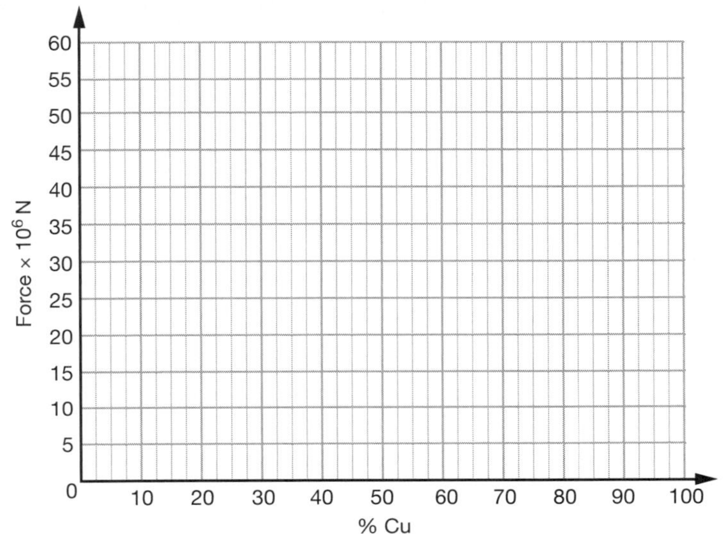

2. Which point on the graph indicates the breaking point of pure copper? Colour the point in red.

3. Which point on the graph indicates the breaking point of pure zinc? Colour the point in green.

4. Use your graph or table to predict the breaking strength of:

   (a) a 50/50 alloy of copper and zinc. Plot this point on your graph. ____________

   (b) an alloy of 25% Cu and 75% Zn ____________

   (c) an alloy containing 25% zinc. ____________

5. Use the graph to identify the proportions of copper and zinc that are needed to make the alloy stronger than pure copper.

2.4 Comparing alloys

6. Identify the proportions of zinc that make the alloy weaker than pure zinc.

____

7. State the percentage of copper that makes the strongest alloy. ____________

8. State the composition of three alloys that all break when a force of 25×10^6 N is applied to them.

9. Conduct research into the metal composition of Australian coins. The coins are made of two different types of alloys. Complete the table identifying the composition of the alloys used to mint the coins. Propose reasons why these alloys are used for minting Australian coins.

mint (coin) (*v*) to produce money by stamping metal

Australian coins and their composition		
Australian coins	5c, 10c, 20c, 50c coins	$1, $2 coins
composition of alloy		
reason for using this alloy		

RATE MY UNDERSTANDING
Shade the face that shows your rating   

2.5 Glass, steel and temperature changes

Science as a human endeavour

FOUNDATION | STANDARD | **ADVANCED**

Materials such as glass and steel develop different properties when heated. Table 2.5.1 describes two processes which glass and metals may be subjected to.

Table 2.5.1

Changing properties of metals and glass with heating		
Name of processes	**Description of processes to treat metals and glass**	**Changed properties of metals or glass**
annealing (or normalising)	Glass and steel are heated and then left to cool naturally.	Glass is toughened by the process. Steel is softened, making it easier to shape into wires and cables.
tempering	Glass and steel are repeatedly heated and then cooled rapidly, usually by being dipped into cold water. This rapid cooling is called quenching—the material has been quenched.	Tempered glass is also known as safety glass because it forms small (and safe) rounded beads when broken. Tempering makes the substances stronger. For centuries, blacksmiths have used tempering to toughen and shape steel tools, horseshoes and the blades of knives and swords.

1 State alternative names for:

(a) annealing ______

(b) tempered glass. ______

2 Define quenching.

______

______

(3) Compare the processes of annealing and tempering using the Venn diagram.

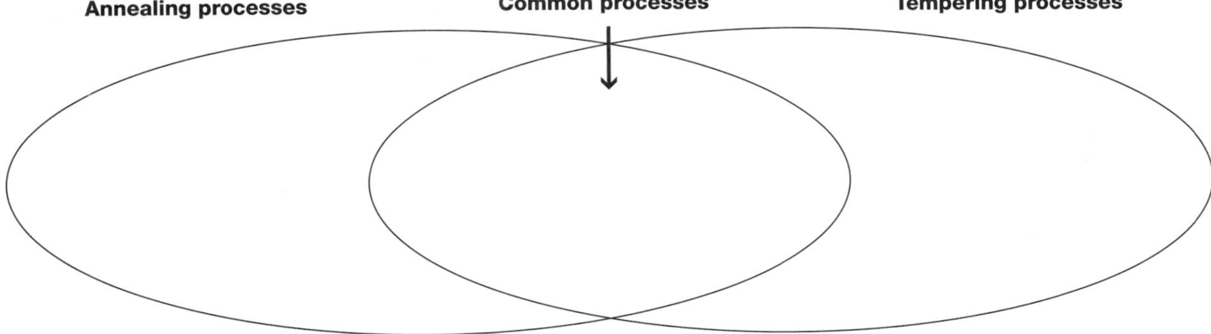

(4) Compare annealed glass with normal glass.

______

______

______

______

blacksmith (*n*) someone who makes things from iron and steel

droop (*v*) to bend downwards

initial (*adj*) first

toughen (*adj*) to make something stronger

withstood (*v*) survived a force or injury

5 Propose a use for:

(a) tempered glass

____

(b) tempered steel.

____

Figure 2.5.2 Debris of one of the collapsed Twin Towers, New York City, USA.

Heat can also cause disaster. When it is hot, steel acts like plastic, and stretches and bends if force is applied to it. Steel structures tend to droop and collapse in an extreme fire. This often happens in factory fires.

The steel structures drooped, causing them to collapse on 11 September 2001, after the Twin Towers of the World Trade Center in New York, United States of America, were struck by aircraft piloted by terrorists. Both towers withstood the initial collisions, but exploding aircraft fuel ignited fires in the buildings. This intense heat caused the steel structures holding up the upper floors to sag and pull in the outer walls of both towers. The weakened walls could not hold the weight of the floors above them and so they collapsed. The impact of the collapsing upper floors then caused the lower walls to collapse again and again until both towers had collapsed.

6 Compare the behaviour of steel at room temperature with steel at extremely high temperature.

____

____

____

____

(7) Aircraft colliding with the World Trade Center towers in New York caused a chain of events that ended in the towers collapsing. Outline how the collision changed the materials in the towers enough to cause them to collapse. Show the changes on the flow diagram.

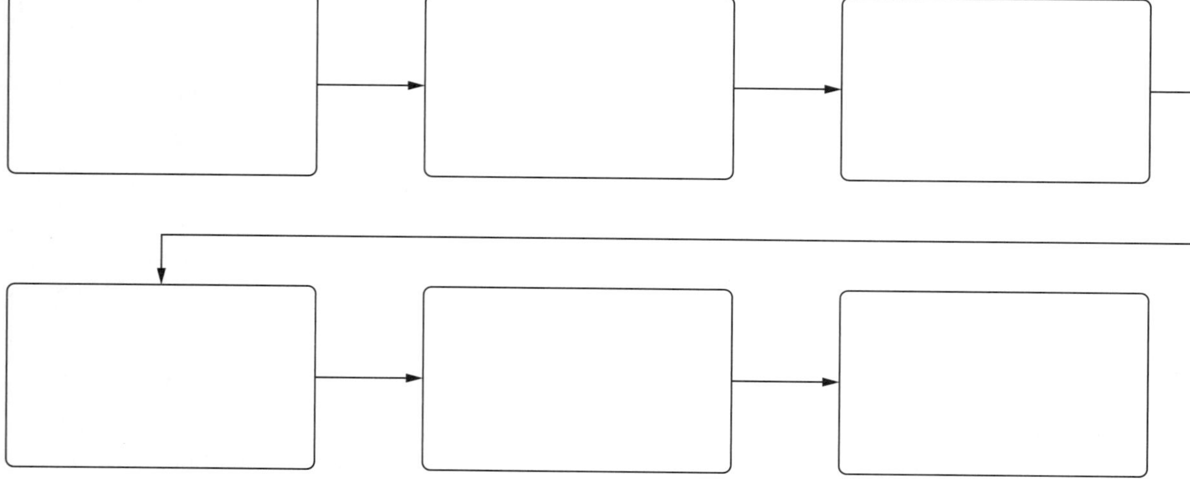

RATE MY UNDERSTANDING
Shade the face that shows your rating

2.6 pH and indicators

Science inquiry skills

FOUNDATION	STANDARD	ADVANCED

Communicating

(1) Different indicators turn different colours at different pH values.

Predict what colour each material would give if tested with the indicators shown. Use your answers to complete the table below.

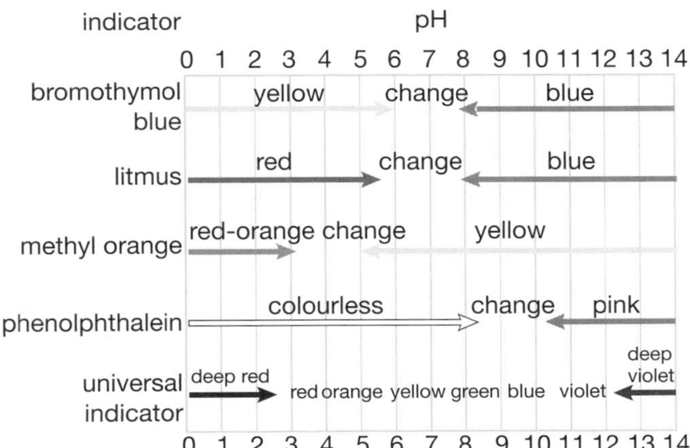

pH indicators

Substance and pH		litmus	bromothymol blue	methyl orange	phenol-phthalein	universal indicator
floor cleaner	10.0					
ammonia solution	11.0					
brass polish	9.5					
calcium hydroxide solution	11.9					
carpet shampoo	5.9					
cream cleanser	8.8					
dilute caustic soda	13.0					
dilute nitric acid	1.0					
dishwashing liquid	5.5					
kitchen cleaner	11.0					
lemon juice	2.5					
milk	6.8					
oranges	3.2					
oven spray	12.5					
tea	5.2					
toothpaste	6.8					
vinegar	2.9					
wine	3.8					

RATE MY UNDERSTANDING
Shade the face that shows your rating

2.7 Literacy review

Science understanding

FOUNDATION | STANDARD | ADVANCED

1 Recall your knowledge of materials to complete this crossword puzzle and the clues. Where clues are provided, write the correct word onto the crossword puzzle. Where words are provided on the crossword puzzle, write the clue for that word.

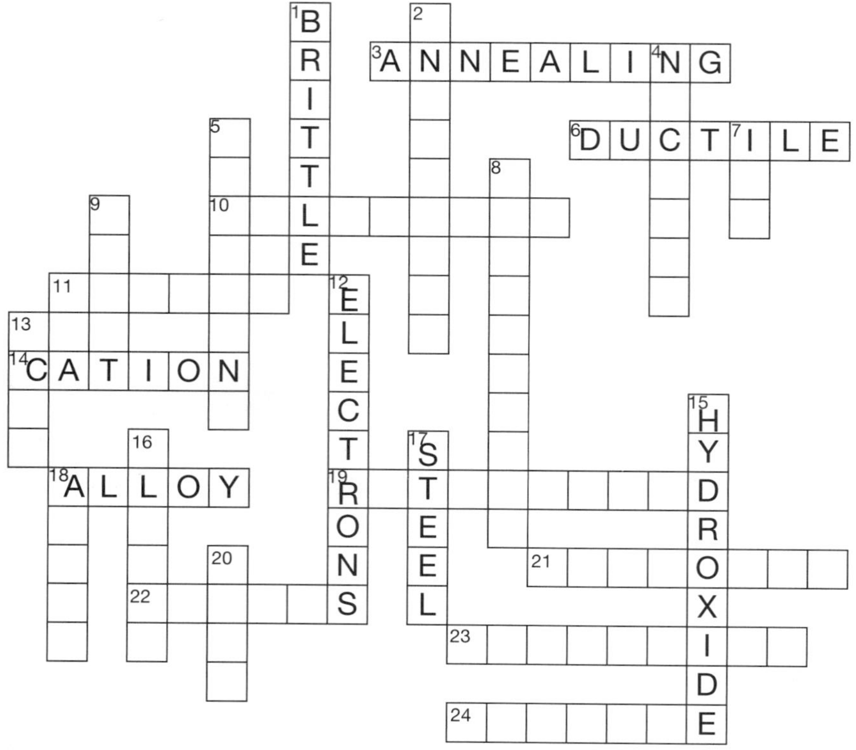

Across

3 ______________________

6 ______________________

10 Able to be hammered into new shapes

11 A positively charged particle in an atom

14 ______________________

18 ______________________

19 Name of the scientist who used gold foil to prove the atom to be largely empty space

21 The table that lists all known elements

22 Turns red in the presence of acid, blue in a base

23 A process of rapidly cooling heated metal by dropping it in water

24 A grid-like structure of atoms or ions

Down

1 ______________________

2 A substance that shows whether another substance is acidic, neutral or basic

4 Core of the atom

5 A pure substance made of two or more different types of atoms that are chemically joined

7 An atom that has gained or lost an electron

8 Different forms of carbon

9 The purity of gold is measured this way

12 ______________________

13 A substance that releases hydrogen ions

15 ______________________

16 A base that dissolves in water

17 ______________________

18 A negative ion

20 Fundamental building block of all materials

RATE MY UNDERSTANDING
Shade the face that shows your rating

2.8 Thinking about my learning

Think back over your learning for this chapter on materials. Your task is to write a report about the work you have done in this topic.

Name:

Class Date

I have been learning about

I have learned how to

When working in groups I am particularly good at

I have helped others with their learning by

My most outstanding piece of work in this topic was

I need to work on improving

CHAPTER 3 Reaction types

3.1 Knowledge preview

Science understanding

FOUNDATION | **STANDARD** | ADVANCED

1 Complete the spider diagram below by writing one thing you know about corrosion in each box.

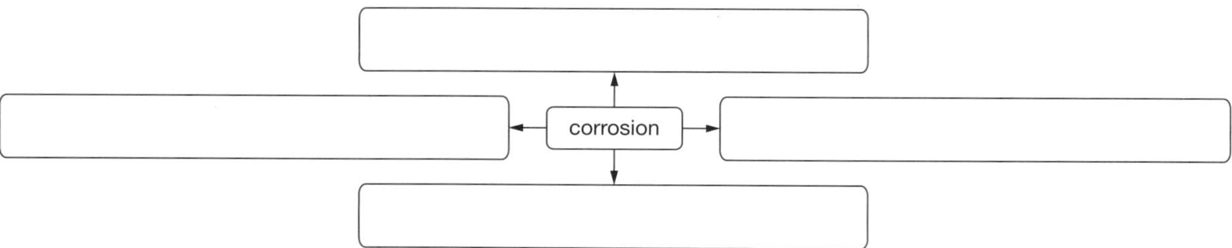

2 Draw a single line to match each word on the left to the chemical formula on the right.

glucose		O_2
carbon dioxide		$C_6H_{12}O_6$
water		CH_4
oxygen		CO_2
methane		H_2O

3 The tables below have statements about photosynthesis and respiration. Analyse the statements and complete the following.

(a) Write a P next to each photosynthesis statement.

(b) Write an R next to every respiration statement.

A chemical reaction requires sunlight to convert water and carbon dioxide into oxygen and glucose.	In this process, oxygen and glucose travel through the bloodstream.
This is the process of obtaining energy from food.	This is the process that plants follow in order to make food.
carbon dioxide + water → oxygen + glucose	oxygen + glucose → carbon dioxide + water
Chlorophyll captures the sunlight that is required for the process.	Carbon dioxide is removed when you breathe out during this process.

3.2 Analysing a reaction

Science inquiry skills

FOUNDATION	**STANDARD**	ADVANCED

Processing & Analysing | Evaluating | Questioning & Predicting

A bright blue solution of copper sulfate ($CuSO_4$) was poured into a beaker containing a colourless solution of barium chloride ($BaCl_2$). A white powder (a precipitate) of barium sulfate ($BaSO_4$) formed immediately. It then settled to the bottom of the beaker. A clear, bright blue solution of copper chloride ($CuCl_2$) remained on top.

The masses of all reactants, products and their beakers were measured. There were four trials in the experiment. The results are shown in the table below.

Reaction	Results of trials with mass in grams (g)			
	Trial 1	Trial 2	Trial 3	Trial 4
Before reaction				
beaker 1 + $BaCl_2$	225.2	225.7	230.1	228.4
beaker 2 + $CuSO_4$	150.1	149.6	149.8	148.1
total				
After reaction				
beaker 1 + $BaSO_4$ + $CuCl_2$	293.8	293.2	299.4	283.3
empty beaker 2	81.5	81.9	80.2	93.2
total				

1. Calculate the total masses before and after each trial. Enter your results in the table above.

2. Analyse your totals and assess whether the results prove or disprove the Law of conservation of mass.

____

3. Justify your answer to question 2.

____

____

4. In Trial 4, the mass of beaker 2 after the reaction was quite different from all the other measurements of this beaker. Analyse the results and propose a reason why.

____

____

5. Construct a word equation for the reaction.

__________ + __________ → __________ + __________

RATE MY UNDERSTANDING
Shade the face that shows your rating

3.3 Balancing chemical equations 1

Science understanding

FOUNDATION	STANDARD	**ADVANCED**

1 Classify the following formula equations as balanced or unbalanced. If unbalanced, identify the elements that are not balanced.

Equation	Balanced/ unbalanced equation	Unbalanced element(s)
(a) $H_2O \rightarrow H_2 + O_2$		
(b) $Fe_2O_3 + 2CO \rightarrow 2Fe + 3CO_2$		
(c) $2K + 2H_2O \rightarrow 2KOH + H_2$		
(d) $2CH + 22O_2 \rightarrow 16CO_2 + 18H_2O$		
(e) $C_6H_{12}O_6 + 6O_2 \rightarrow 6CO_2 + 6H_2O$		
(f) $C_2H_6 + 7O_2 \rightarrow 2CO_2 + 3H_2O$		
(g) $2CH + 3O \rightarrow 2CO + 4H_2O$		
(h) $2Al + 3O_2 \rightarrow 4Al_2O_3$		
(i) $CaCO_3 + H_2SO_4 \rightarrow CaSO_4 + H_2O + CO_2$		
(j) $CH_3COOH + NaHCO_3 \rightarrow NaCH_3COO + H_2O + CO_2$		

2 Determine the missing number that will balance the following formula equations.

(a) ________ $HCl + CaCO_3 \rightarrow CaCl_2 + H_2O + CO_2$

(b) $CH_4 +$ ________ $O_2 \rightarrow CO_2 +$ ________ H_2O

(c) $C_3H_8 +$ ________ $O_2 \rightarrow 3CO_2 +$ ________ H_2O

(d) $Ca +$ ________ $HCl \rightarrow CaCl_2 + H_2$

(e) $4Fe +$ ________ $O_2 \rightarrow$ ________ Fe_2O_3

3 Modify the following equations to make them balanced. Note that some of them might already be balanced.

(a) $H_2SO_4 + NaOH \rightarrow Na_2SO_4 + H_2O$

(b) $Mg + O_2 \rightarrow MgO$

(c) $H_2O + CO_2 \rightarrow H_2CO_3$

(d) $Ag + H_2S \rightarrow Ag_2S + H_2$

(e) $Na + H_2O \rightarrow NaOH + H_2$

(f) $HCl + Mg \rightarrow MgCl_2 + H_2$

(g) $H_2SO_4 + Fe \rightarrow FeSO_4 + H_2$

(h) $HNO_3 + K_2O \rightarrow KNO_3 + H_2O$

(i) $HCl + CaCO_3 \rightarrow CaCl_2 + H_2O + CO_2$

(j) $NO_2 + H_2O \rightarrow HNO_2 + HNO_3$

(k) $SO_2 + H_2O + O_2 \rightarrow H_2SO_4$

4 Identify the elements present in each of the following equations.

(a) State how many atoms of each element are present in the reactants and the products. The first question is done for you.

(b) Is the equation is balanced (B) or unbalanced (U)? Circle the correct alternative.

Equation elements and atoms	Elements	No. of atoms in reactants	No. of atoms in products	Balanced or unbalanced?	
(i) $C_7H_{16} + O_2 \rightarrow CO_2 + H_2O$	carbon hydrogen oxygen	7 16 2	1 3 2	B	(U)
(ii) $C_3H_8 + 5O_2 \rightarrow 3CO_2 + 4H_2O$				B	U
(iii) $FeCl_3 + 3NaOH \rightarrow Fe(OH)_3 + 3NaCl$				B	U
(iv) $Mg(OH)_2 + 2HCl \rightarrow MgCl_2 + H_2O$				B	U
(v) $Ca(OH)_2 + H_3PO_4 \rightarrow Ca_3(PO_4)_2 + H_2O$				B	U

5 For each equation in question 4 that you identified as unbalanced, rewrite the balanced equation.

____

____

____

____

____

____

RATE MY UNDERSTANDING Shade the face that shows your rating			

3.4 Balancing chemical equations 2

Science understanding

FOUNDATION | STANDARD | **ADVANCED**

1. Balance the following equations.

(a) _____ H_2SO_4 + _____ HI → _____ H_2S + _____ I_2 + _____ H_2O

(b) _____ FeS_2 + _____ O_2 → _____ Fe_2O_3 + _____ SO_2

(c) _____ Na_2CO_3 + _____ HCl → _____ NaCl + _____ CO_2 + _____ H_2O

(d) _____ $KClO_3$ → _____ KCl + _____ O_2

(e) _____ NaOH + _____ H_2CO_3 → _____ Na_2CO_3 + _____ H_2O

(f) _____ Li + _____ $AlCl_3$ → _____ LiCl + _____ Al

(g) _____ SiF_4 + _____ H_2O → _____ H_4SiO_4 + _____ H_2SiF_6

(h) _____ $Fe_2(SO_4)_2$ + _____ KOH → _____ K_2SO_4 + _____ $Fe(OH)_2$

(i) _____ $Al(OH)_3$ + _____ H_2CO_3 → _____ $Al_2(CO_3)_2$ + _____ H_2O

(j) _____ $NaHCO_3$ + _____ H_2SO_4 → _____ Na_2SO_4 + _____ CO_2 + _____ H_2O

(k) _____ $Pb(NO_3)_2$ → _____ PbO + _____ NO_2 + _____ O_2

(l) _____ $SrBr_2$ + _____ $(NH_4)_2CO_3$ → _____ $SrCO_3$ + _____ NH_4Br

(m) _____ $(NH_4)_2Cr_2O_7$ → _____ NH_3 + _____ Cr_2O_3 + _____ O_2 + _____ H_2O

(n) _____ $Fe_2(C_2O_4)_3$ → _____ FeC_2O_4 + _____ CO_2

(o) _____ MnO_2 + _____ HCl → _____ $MnCl_2$ + Cl_2 + _____ H_2O

(p) _____ $CoBr_3$ + _____ $CaSO_4$ → _____ $CaBr_2$ + _____ $Co_2(SO_4)_3$

(q) _____ NaCN + _____ $CuCO_3$ → _____ Na_2CO_3 + _____ $Cu(CN)_2$

(r) _____ BaO + _____ HOCl → _____ $BaCl_2$ + _____ O_2 + _____ H_2O

(s) _____ $Ca(PO_4)_2$ + _____ SiO_2 + _____ C → _____ P_4 + _____ CO + _____ $CaSiO_3$

(t) _____ Ba_3N_2 + _____ H_2O → _____ $Ba(OH)_2$ + _____ NH_3

(u) _____ $TiCl_4$ + _____ H_2O → _____ TiO_2 + _____ HCl

(v) _____ $Hg(OH)_2$ + _____ H_3PO_4 → _____ $Hg_3(PO_4)_2$ + _____ H_2O

(w) _____ C_2H_6O + _____ O_2 → _____ CO_2 + _____ H_2O

(x) _____ $BaCl_2$ + _____ Na_2SO_4 → _____ NaCl + _____ $BaSO_4$

(y) _____ $Pb(NO_3)_2$ + _____ H_2SO_4 → _____ $PBSO_4$ + _____ HNO_3

(z) _____ C_2H_5OH + _____ O_2 → _____ CO_2 + _____ H_2O

RATE MY UNDERSTANDING
Shade the face that shows your rating

3.5 Sulfuric acid

Science inquiry skills

FOUNDATION	STANDARD	ADVANCED

Sulfuric acid (chemical formula H_2SO_4) is a strong acid. A concentrated solution of sulfuric acid has a pH of 2. Sulfuric acid is an important chemical in industry because it is cheap and used in the production of many other common substances. Sulfuric acid is used in the production of:

- fertilisers (superphosphate and ammonium sulfate)
- anaesthetic (ether)
- explosives like nitroglycerin
- detergents, soaps and plastics.

Sulfuric acid is classified as a desiccant. This means it extracts water from whatever it comes into contact with. For example, you may have seen a demonstration in which concentrated sulfuric acid is added to table sugar (sucrose, chemical formula $C_{12}H_{22}O_{11}$). The sulfuric acid extracts water from the sugar, releasing it as steam, leaving a pile of black carbon. Although the reaction goes through a number of stages, the change to sugar can be summarised as:

$$C_{12}H_{22}O_{11} \xrightarrow{H_2SO_4} 12C + 11H_2O$$

1 What is the chemical formula of sulfuric acid? ______________________

2 What is the pH of concentrated sulfuric acid? ______________________

3 Define the term *desiccant*. ______________________

____

4 What is the chemical formula of table sugar (sucrose)?

____

5 Describe what sulfuric acid does to sugar, by writing its:

(a) word equation ______________________

(b) balanced formula equation ______________________

6 The percentage of sulfuric acid used in the production of different substances is shown in Table 3.5.1.

Table 3.5.1

Percentage of sulfuric acid in different substances							
	Detergents/ soaps	Dyes/ pigments	Fertilisers	Fibres/ plastics	Manufacturing other chemicals	Other uses	Total
% sulfuric acid used	13	24	10	12	26	15	100

Use the information in the table to construct a column graph to show the different uses of sulfuric acid. Label the missing percentages on the vertical axis and label the missing substances on the horizontal axis. Each column should be shown in a different colour.

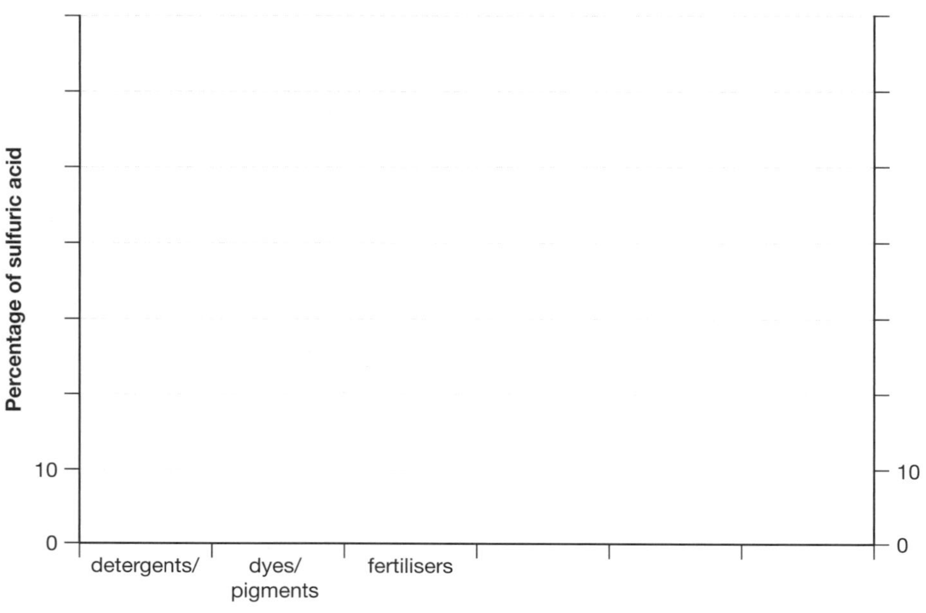

7 Read each statement and indicate whether it is true, false or you cannot tell by looking at the graph.

Statements about sulfuric acid	True / False / Cannot tell
Of all the sulfuric used, the largest percentage is to make dyes and pigments.	
The biggest proportion of sulfuric acid is used to make other chemicals.	
Twice as much sulfuric acid is used to make fertilisers than to make explosives.	
The manufacture of substances containing sulfuric acid is safe as it is not a hazardous chemical.	
Over half of the sulfuric acid used in manufacturing substances goes into detergents, soaps, dyes, pigments and fertilisers.	

RATE MY UNDERSTANDING
Shade the face that shows your rating

3.6 Acid rain

Science understanding

FOUNDATION	**STANDARD**	ADVANCED

Consider the causes and environmental effects of acid rain to complete these questions. You may find it useful to conduct research on this topic.

1 What three chemicals cause rainwater to turn into acid rain? Name the chemicals and write the chemical formula of each.

____

____

____

(2) Rainwater is naturally acidic. Write a word equation and/or balanced formula equation to demonstrate why rainwater is naturally acidic.

____

____

(3) Write balanced formula equations for the following reactions to demonstrate how acid rain forms.

(a) sulfur dioxide + water vapour → sulfurous acid

____

(b) sulfur dioxide + water + oxygen gas → sulfuric acid

____

(c) nitrogen dioxide + water → nitrous acid + nitric acid

____

(4) The chemical equation below shows what happens to marble when acid rain strikes it.

$$CaCO_3 + H_2SO_4 \rightarrow CaSO_4 + H_2O + CO_2$$

(a) Write a word equation for this reaction.

____

____

(b) Use this equation to explain why marble statues are losing their detail because of acid rain.

____

____

RATE MY UNDERSTANDING
Shade the face that shows your rating

3.7 Rate of photosynthesis

Science understanding

FOUNDATION | **STANDARD** | ADVANCED

diffuse (*v*) to move in or out due to the natural motion of the particles

limit (*v*) to control, to restrict

As the chemical equations below show, carbon dioxide and water have to be present before photosynthesis can take place.

carbon dioxide + water → glucose + oxygen

$$6CO_2 + 6H_2O \rightarrow C_6H_{12}O_6 + 6O_2$$

Chlorophyll needs to be present, although it is not a reactant. Sunlight is also needed, because it powers the reaction. If any one of these is not available, then photosynthesis slows down or stops. The factor that is missing or in short supply is called the limiting factor. In times of drought, water could be a limiting factor. If the stomata are closed to save water, then carbon dioxide cannot enter the leaf, and carbon dioxide could become the limiting factor.

At night, light is the limiting factor, so there is no photosynthesis. As the Sun rises, the light intensity increases, so the rate of photosynthesis increases. Then, at some point, the rate of photosynthesis does not increase any further. Something other than light is controlling the rate. When light is no longer the limiting factor, it is often the amount of carbon dioxide in the atmosphere, and the rate at which it can diffuse into the leaf, that limits the rate of photosynthesis. This is shown in Figure 3.7.1.

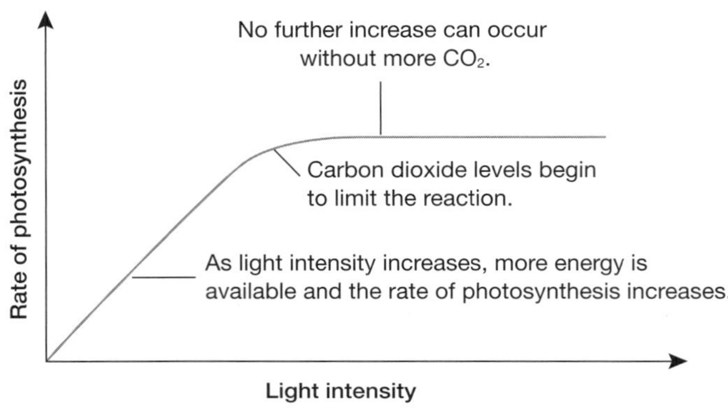

Figure 3.7.1 Rate of photosynthesis as light intensity varies

The speed of most chemical reactions increases as the temperature increases. This is true of photosynthesis too, but only up to a point. The rate of photosynthesis increases until the temperature reaches about 30°C. At this temperature, photosynthesis decreases and then stops. This is because the enzymes that help photosynthesis along are destroyed at temperatures greater than 30°C. Above 30°C the enzymes can no longer function.

1 For the photosynthesis reaction, list the names and chemical formulas of the:

(a) reactants ______________________________

(b) products. ______________________________

2 State what else is required for photosynthesis to occur.

____

3.7 Rate of photosynthesis

3 Write the word equation and the balanced formula for photosynthesis.

(a) word equation ______

(b) balanced chemical formula. ______

4 Identify the limiting factor that stops photosynthesis from proceeding at night.

______

5 Plants don't grow well in a drought. They may even die. Explain why this happens.

______

6 Most caves have no plants in them. Identify the limiting factors of this environment that would make it difficult for plants to grow there.

7 **(a)** State the temperature above which photosynthesis stops. ______

(b) Explain why photosynthesis stops above this temperature.

______

8 Compare lines X and Y on the graph shown below.

Amount of oxygen production during photosynthesis with various light intensities

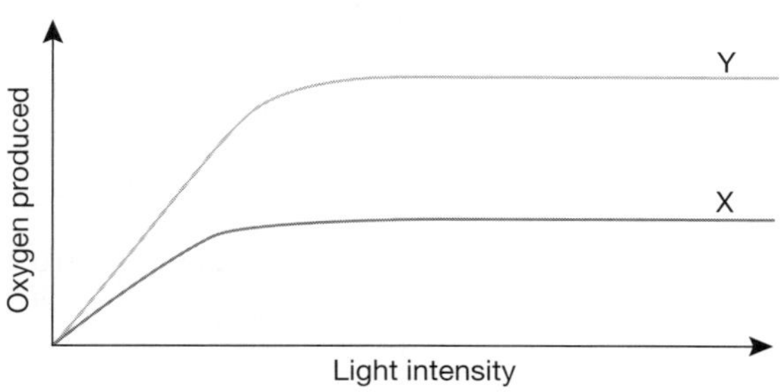

9 Summarise what both lines on the graph show about photosynthesis as light intensity increases.

______

10 Discuss possible reasons for the difference in the lines.

______

______

11 Explain why the amount of oxygen produced is a good measure of the rate of photosynthesis taking place.

______

______

RATE MY UNDERSTANDING Shade the face that shows your rating	☺	😐	☹

3.8 Anaerobic respiration

Science understanding

FOUNDATION | **STANDARD** | ADVANCED

Aerobic respiration is the way animals get most of their energy. The muscles of animals usually have lots of oxygen. However, there are times when oxygen is used at a faster rate than the rate of supply. The cells of the muscles then have to get their energy by respiring without oxygen. Respiration without oxygen is called anaerobic respiration.

Glucose is still the source of the energy, but lactic acid ($C_3H_6O_3$) is produced instead of carbon dioxide and water.

Anaerobic respiration in animals can be represented by the chemical equations:

glucose → lactic acid + energy

$C_6H_{12}O_6 \rightarrow 2C_3H_6O_3 + \text{energy}$

Anaerobic respiration does not produce as much energy as aerobic respiration because some energy is still stored in the lactic acid molecule.

After an extended period of exercise, a build-up of lactic acid in the muscles can cause muscle fatigue. Warming down and breathing deeply helps supply oxygen to the muscles. Oxygen helps the lactic acid break down, releasing more energy.

1 For anaerobic respiration in animals, name the:

(a) reactant ______________________________

(b) waste product. ______________________________

2 Name the chemical that can cause muscle soreness.

____

____

(3) Explain why you don't feel sore normally, even though you are moving about and doing constant light exercise.

____

____

(4) **(a)** Recall sports or activities after which your muscles felt sore.

____

____

(b) Explain what was probably produced when your muscles felt sore and why.

____

____

3.8 Anaerobic respiration

5 Compare aerobic respiration with anaerobic respiration for both plants and animals by listing their similarities and differences in the Venn diagram below.

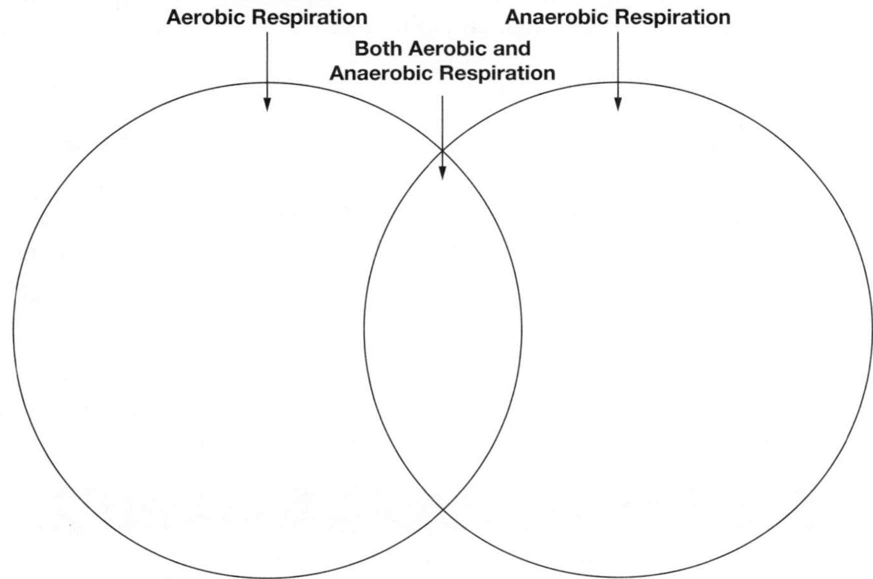

6 Jan was running in a 400-metre race. Just before the finish line she developed cramps in her legs caused by severe muscle fatigue. She fell down, and as she lay there taking deep breaths, the cramps eased. Analyse what was happening to Jan. Explain why these events occurred.

7 Bacteria can be classified as aerobic or anaerobic. How do the conditions for each type of bacteria differ?

Conditions for aerobic bacteria	Conditions for anaerobic bacteria
Aerobic bacteria are exposed to the air, so …	Anaerobic bacteria are not exposed to the air, so …

8 Suggest reasons why high-powered, energetic gym classes are sometimes known as aerobics, and not anaerobics.

RATE MY UNDERSTANDING
Shade the face that shows your rating

3.9 Half-life decay

Science understanding

FOUNDATION | **STANDARD** | ADVANCED

Jonathon is measuring the decay of 1000 polonium-218 atoms. He knows that polonium-218 decays rapidly into lead-214 through alpha decay.

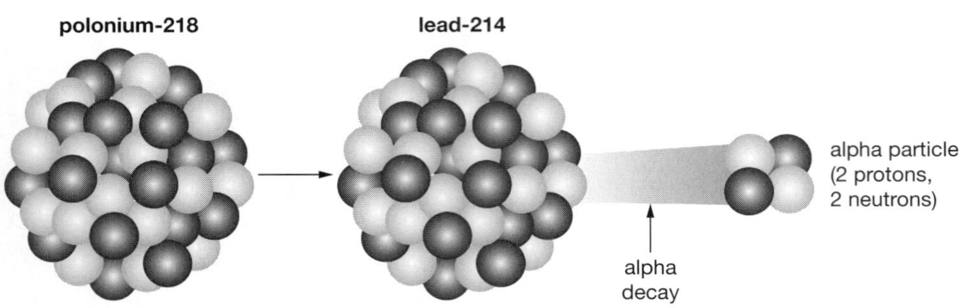

decay (*v*) the emission of energetic particles from a nucleus

half-life (*n*) the time it takes a substance to decrease by half

line of best fit (*n*) a line on a graph that fits most closely to the data

To measure the half-life of polonium-218, Jonathon counted the number of polonium-218 atoms he had left after every 2 minutes. His results are written in the table below.

Decay of polonium measured over time											
Time (min)	0	2	4	6	8	10	12	14	16	18	20
Number of polonium-218 atoms	1000	635	430	280	155	85	65	35	25	15	10

(1) Plot the data in the table on the axes provided. Plot only the data points, but no connecting lines.

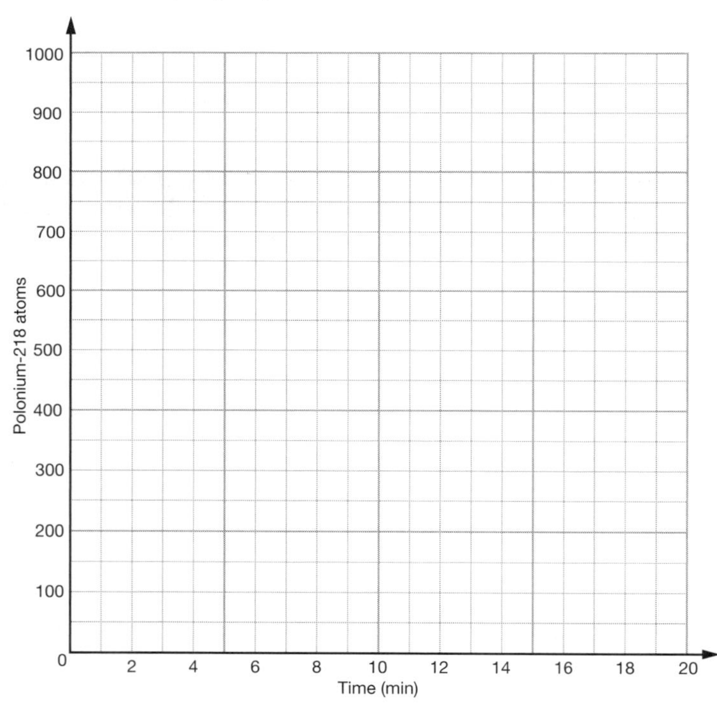

2. To show the general trend in the decay, draw a curve of best fit though the data points.

3. State how many polonium-218 atoms Jonathon has after:

   (a) 4 min ____________________

   (b) 14 min ____________________

   (c) 9 min ____________________

   (d) 19 min. ____________________

4. Determine the time (min) when Jonathon has:

   (a) one-half of the original number of polonium-218 atoms

   (b) one-quarter of the original number of polonium-218 atoms

   (c) one-eighth of the original number of polonium-218 atoms

   (d) one-sixteenth of the original number of polonium-218 atoms.

5. Calculate the half-life of polonium-218.

3.10 Isotopes

Science understanding

FOUNDATION | **STANDARD** | ADVANCED

The atomic number is the number of protons in the nucleus of an atom. The number of protons tells us the type of atom it is. For example, all oxygen atoms have 8 protons, all chlorine atoms have 17 protons and all gold atoms have 79 protons.

The mass number is the number of protons plus neutrons in the nucleus. Atoms with the same atomic number but different mass numbers are known as isotopes. For example, carbon atoms have 6 protons but may have 6, 7 or 8 neutrons. This gives three isotopes: carbon-12, carbon-13 and carbon-14. The number 12, 13 or 14 refers to the mass number.

Zoe has a 65 g block of zinc metal. She knows that zinc has an atomic number of 30 and five stable isotopes: zinc-64, zinc-66, zinc-67, zinc-68 and zinc-70.

1 Calculate the number of neutrons in each isotope.

Zinc-64: ______________________________

Zinc-66: ______________________________

Zinc-67: ______________________________

Zinc-68: ______________________________

Zinc-70: ______________________________

2 Each type of atom may have several isotopes but only some isotopes are radioactive. Radioactive decay is the process whereby an unstable atom emits radiation and loses energy. There are three main types of radiation that can be emitted: alpha, beta and gamma. Use the words in the box below to complete the sentences:

alpha	beta	decay	ejected	electrons	gamma
isotopes	mass number	negative	neutrons	nucleus	powerful
protons	radiation	ray	rearrange	small	wave

______________ are atoms that have the same number of protons but a different number of ______________. This also means there is a change in the ______________. Radioactive ______________ is the process by which an unstable atom emits ______________. During ______________ decay, a ______________ ejects an alpha particle which is a cluster of 2 ______________ and 2 neutrons. During beta decay, a nucleus ejects a ______________ particle. Beta particles are identical to ______________ as they are very ______________ and have a ______________ charge. The last type of radiative decay is ______________. In this type of decay, no particles are ______________ from the nucleus but the protons and electrons ______________ themselves inside of the nucleus. They emit a ______________ electromagnetic ______________ called a gamma ______________.

RATE MY UNDERSTANDING
Shade the face that shows your rating

3.11 Literacy review

Science understanding

FOUNDATION | STANDARD | ADVANCED

1 Match the jumbled list of terms in the left column with their correct definitions in the right column. Match them by writing the correct term in the middle column next to its correct definition. There are some terms and definitions missing from the table. Write them in.

Terms and definition match

Jumbled terms	Correct terms	Definitions
stomata		chemicals that take part in a chemical reaction
oesophagus		
salt		the simple sugar made by photosynthesis
conservation		a reaction that absorbs energy
aerobic		chemicals produced in a chemical reaction
		the correct name for your food pipe
endothermic		single-celled organism that lives in water
corrosion		formed when a metal takes the place of a hydrogen atom in an acid
reactants		a rapid reaction with oxygen, releasing heat and light
carbonates		mass obeys this law
diatom		pain caused by stomach acid rising into your windpipe
exothermic		
glucose		rust is an example of this reaction
neutralisation		how cells get their energy
combustion		the green chemical needed for photosynthesis
products		a reaction between an acid and a base
enzymes		the common name for hydrated iron(III) oxide
chlorophyll		reaction of an acid with these chemicals produces carbon dioxide
respiration		a reaction that releases energy
		these are biological catalysts

RATE MY UNDERSTANDING
Shade the face that shows your rating

3.12 Thinking about my learning

1 Reflect on what you have learned in this chapter on reaction types. With a partner, write down what you have learnt in the circle on the left. Your partner writes what he/she has learnt in the circle on the right. Discuss and compare your learning.

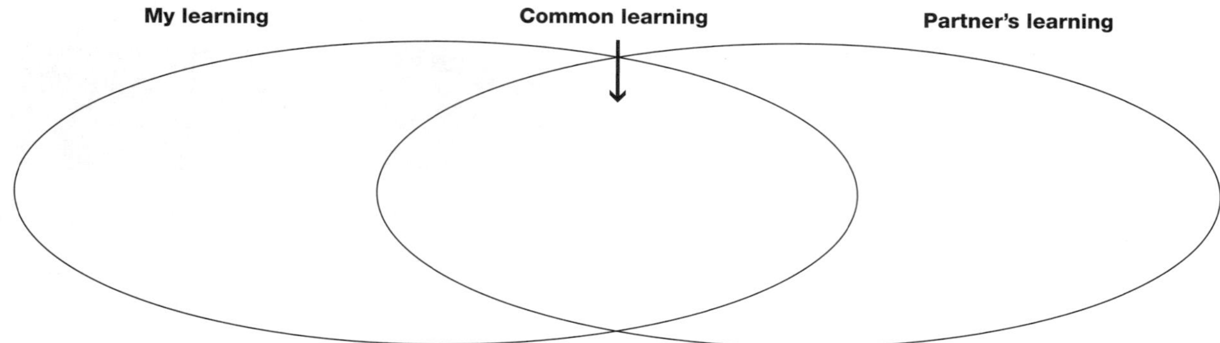

2 **(a)** Select any three of the things you learnt about reaction types that you wrote on the Venn diagram. Create three open-ended questions based on these areas of the chapter. An open-ended question is one that requires a detailed answer, not just a one word answer.

Question 1 ______________________________

Question 2 ______________________________

Question 3 ______________________________

(b) Write detailed answers for each of your questions. Try to do this without referring back to any information.

Answer to Question 1 ______________________________

Answer to Question 2 ______________________________

Answer to Question 3 ______________________________

(c) Test each other's understanding by taking it in turns to ask each other the questions you wrote and check the answers.

CHAPTER 4

Heat, light and sound

4.1 Knowledge preview

Science understanding

FOUNDATION | **STANDARD** | ADVANCED

1 Draw a single line to match up each energy type in the left column with its definition in the right column.

Energy type
elastic potential energy
chemical energy
kinetic energy

Energy definition
the energy an object possesses due to its movement or motion
the energy that is stored as a result of an object's stretching or compressing
an energy that is stored in food

(2) Heat, light and sound are all types of energy. Energy can never be destroyed, but it can be transformed (changed) into a different type of energy or transferred from one object to another. If energy is transformed into different types of energy, these changes can be represented in an energy flow diagram.

(a) In the boxes on the left of the energy flow diagrams, state the types of energy that power the TV and the car.

(b) In the boxes on the right, state what types of energy the initial energy has been transformed into. This could be more than one type of energy.

(c) Converted energy can be either useful or wasted. Useful energy is the desired energy. Wasted energy is a by-product and was not intentionally created. On each of the energy flow charts, highlight the useful energies in green and the wasted energies in red.

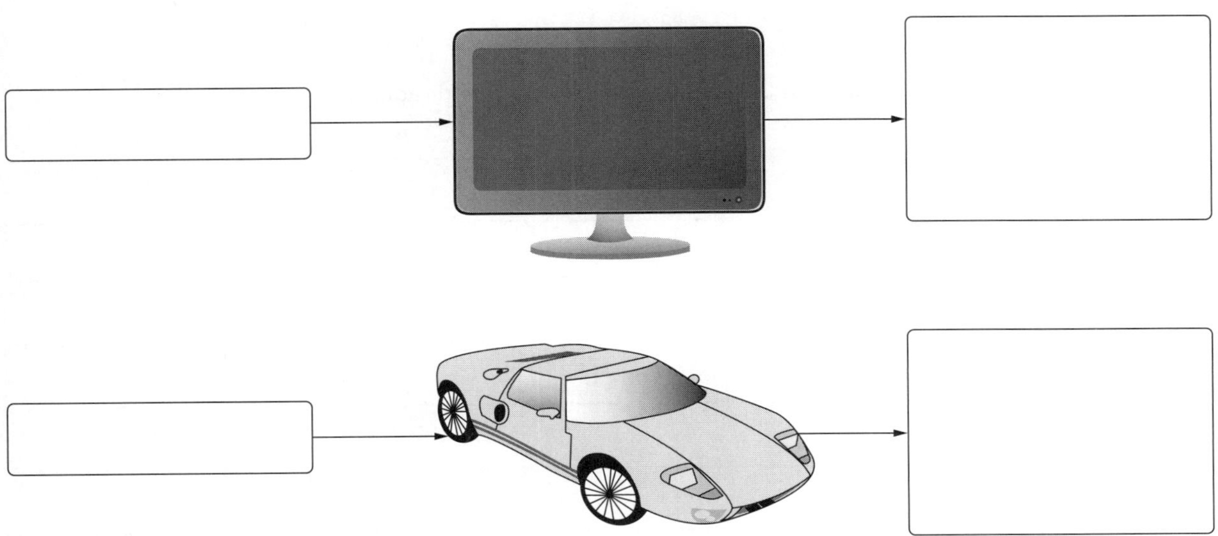

4.1 Knowledge preview

3 In the box provided, write down everything you know about the conduction, convention and radiation of heat.

Conduction	Convection	Radiation

4 Using the diagram below, compare and contrast conduction, convection and radiation. This means finding the similarities and differences between them. To challenge yourself, try to refer to particles.

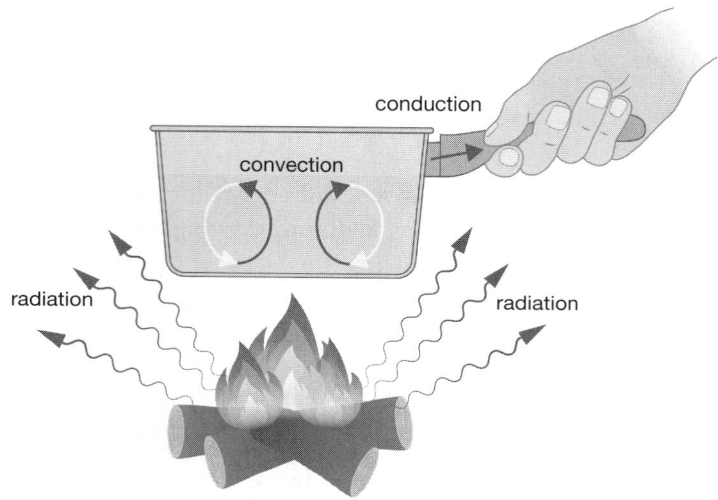

Figure 4.1.3 Conduction, convection and radiation

4.2 Testing insulators

Science inquiry skills

FOUNDATION | **STANDARD** | ADVANCED

Processing & Analysing **Communicating**

Ethan and Ashia designed an experiment to test how well three materials insulate against heat loss. They tested cotton wool, paper scraps and pieces of foam. Ethan and Ashia set up equipment as shown in Figure 4.2.1.

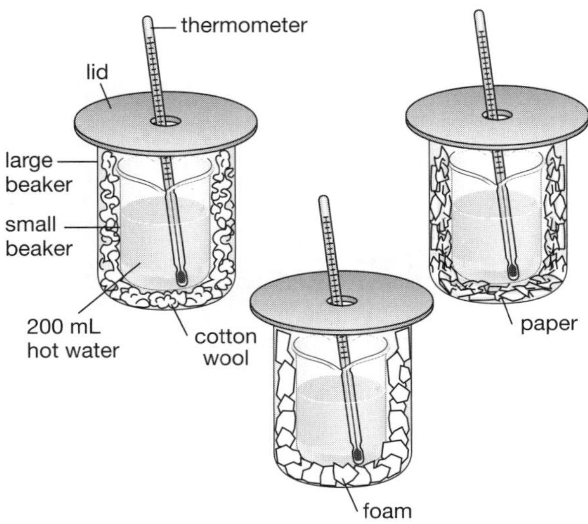

Figure 4.2.1 Materials insulation test

They poured 200 mL of boiling water into each beaker and then recorded the water temperature every minute for 15 minutes. Their results are shown in the table below.

Time (min)	Water temperature (°C)		
	Cotton wool	**Foam**	**Paper**
0	97	96	94
1	80	82	77
2	66	72	59
3	54	65	47
4	48	60	43
5	44	57	37
6	41	54	36
7	40	52	35
8	39	50	34
9	38	48	34
10	37	48	33
11	36.5	47	32.5
12	36.5	46.5	32.5
13	36	46.5	32
14	36	46	32
15	36	46	32

4.2 Testing insulators

Construct three line graphs, one for each insulator, showing how the water temperature dropped for each sample tested. Use the same set of axes as shown in Figure 4.2.2. Label each line graph as cotton wool, foam or paper.

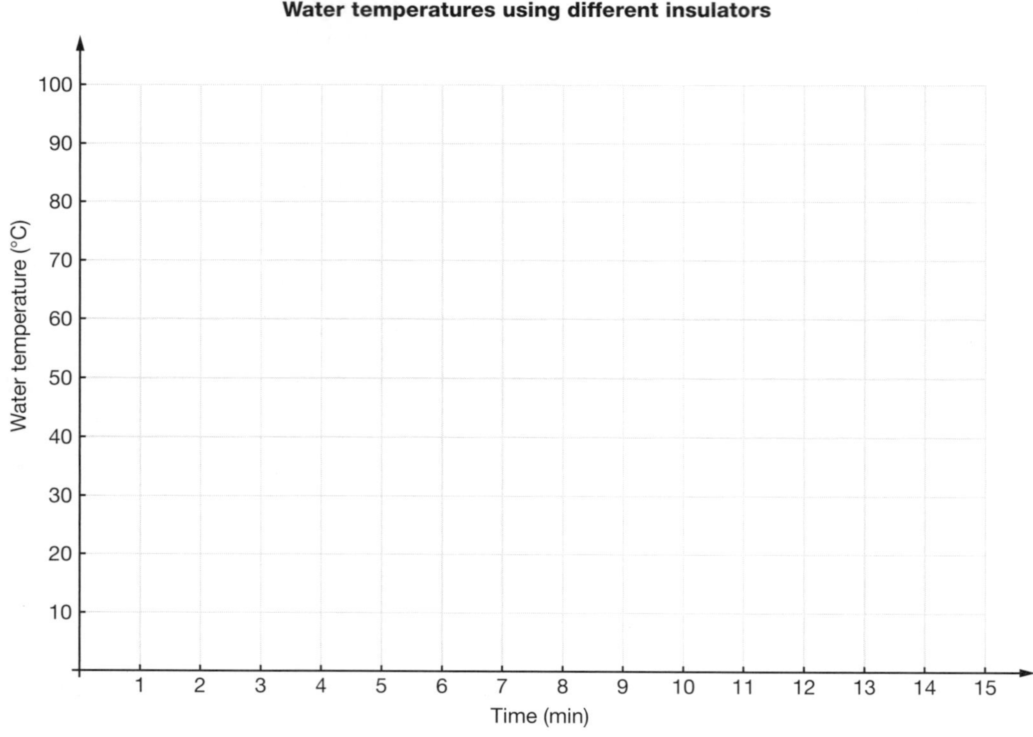

Figure 4.2.2

(1) Rank the three materials tested, in order, from best insulator to poorest insulator.

____

2 **(a)** Explain why the water temperature didn't start from 100°C.

____

(b) Explain why the starting temperatures were different for each of the beakers.

____

(3) List three variables. For each variable, explain how it was controlled in Ethan and Ashia's experiment.

1 ______________________________

2 ______________________________

3 ______________________________

(4) Name four places where we use insulating materials.

1 ______________________________

2 ______________________________

3 ______________________________

4 ______________________________

RATE MY UNDERSTANDING
Shade the face that shows your rating

4.3 Cool cars

Science as a human endeavour

FOUNDATION | STANDARD | **ADVANCED**

The California Global Warming Solutions Act (2006) aims to reduce California's greenhouse emissions to 1990 levels by the year 2020. One proposal suggested that people should reduce the amount of heat absorbed by their cars when parked in the sun. If less heat was absorbed by cars, then people would use their car airconditioners less, reducing greenhouse gas emissions. The US Department of Energy conducted a series of experiments (called the IMAC Vehicle Soak Test) to determine whether three types of changes made to a car reduced the amount of heat energy absorbed by a car. These changes were:

circulate (*v*) to move around

greenhouse (gas) emissions (*n*) gases such as CO_2 which cause the atmosphere to trap in heat

modify (*v*) to change

- using solar reflective paint that reflects at least 20% of solar energy
- installing a solar panel in the sunroof of the car, and using this to power six small fans to circulate air inside the car
- using solar reflective window glazing to reflect more solar energy, with the windshield reflecting at least 30% of solar energy.

The experiments were conducted using a Cadillac STS, which was modified with the three changes listed above, and using an identical car with no modifications. The cars were parked next to each other, and temperatures at seven positions within the cars (Figure 4.3.1) were measured over a number of hours and on different days. Averages of these temperatures were used to calculate the temperature differences between the two cars. Temperatures were lower in every position in the modified car. Figure 4.3.2 shows a graph of the average reduction in temperatures measured at the seven positions in the cars.

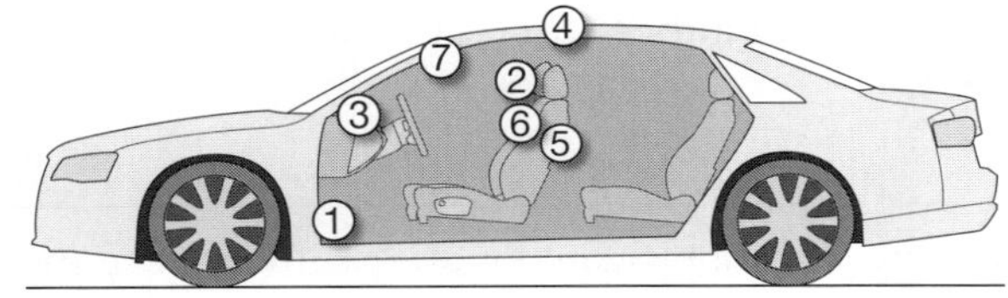

Figure 4.3.1

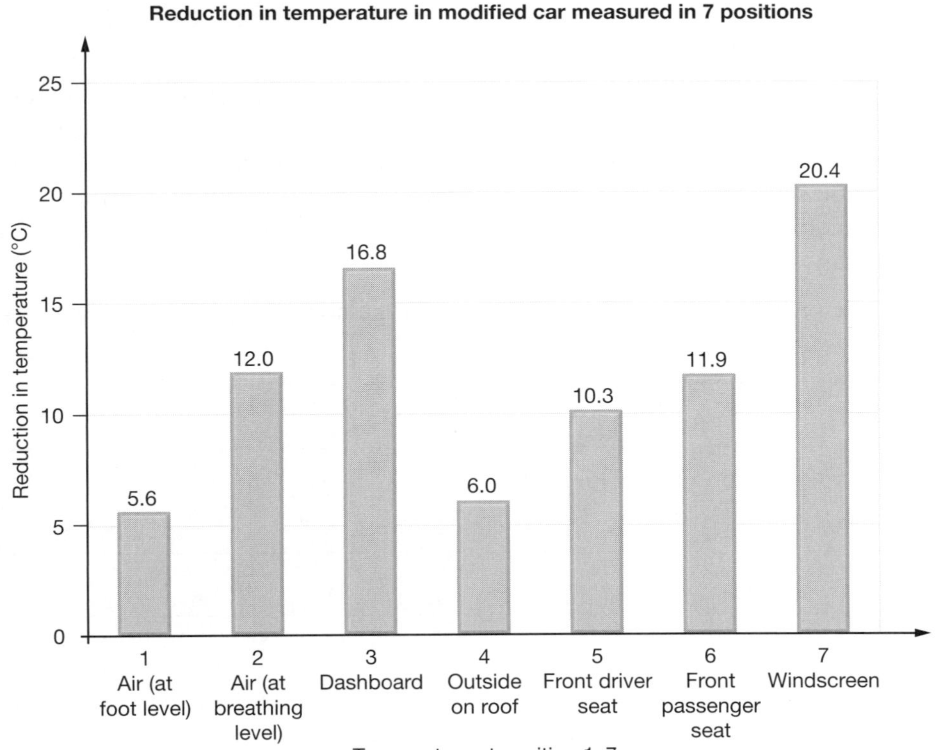

Figure 4.3.2 7 positions for temperature measurement in a car

4.3 Cool cars

1. State which paint colours would absorb more infrared radiation.

2. Which colour/s of car do you think would be the coolest?

3. Describe how you think each of the changes made would reduce the heat absorbed by the car using:

   (a) solar-reflective paint

   (b) solar-powered fans to circulate air inside the car

   (c) solar-reflective window glazing.

4. Use figures 4.3.1 and 4.3.2 to answer the following questions.

   (a) Which position in the modified car recorded the greatest temperature drop?

   (b) What was the reduction in temperature at the driver's head level (or breathing level) in the modified car?

   (c) Do you think these modifications would make a difference in the use of car airconditioners?

5. After making these proposals, it was found that no black automotive paint could be made to reflect 20% of solar energy. The closest possible colour was described as 'mud puddle brown'. White is the most popular car colour in the USA. Black is the second most popular car colour, and so many people were angry that these recommendations could be made into a law that would ban black cars. As a result, the Californian government is only regulating that car windows must have reflective glazing.

   (a) State one reason for and one reason against regulating the paint colour of cars.

   (b) What is your opinion on this issue?

RATE MY UNDERSTANDING
Shade the face that shows your rating

4.4 The ear

Science understanding

FOUNDATION	STANDARD	ADVANCED

Demonstrate your understanding of the structure of the ear by labelling the diagram in Figure 4.4.1 using the following terms:

anvil	**auditory nerve**	**cochlea**	ear canal
eardrum	Eustachian tube	hammer	**ossicles**
oval window	pinna	**semicircular canals**	stirrup

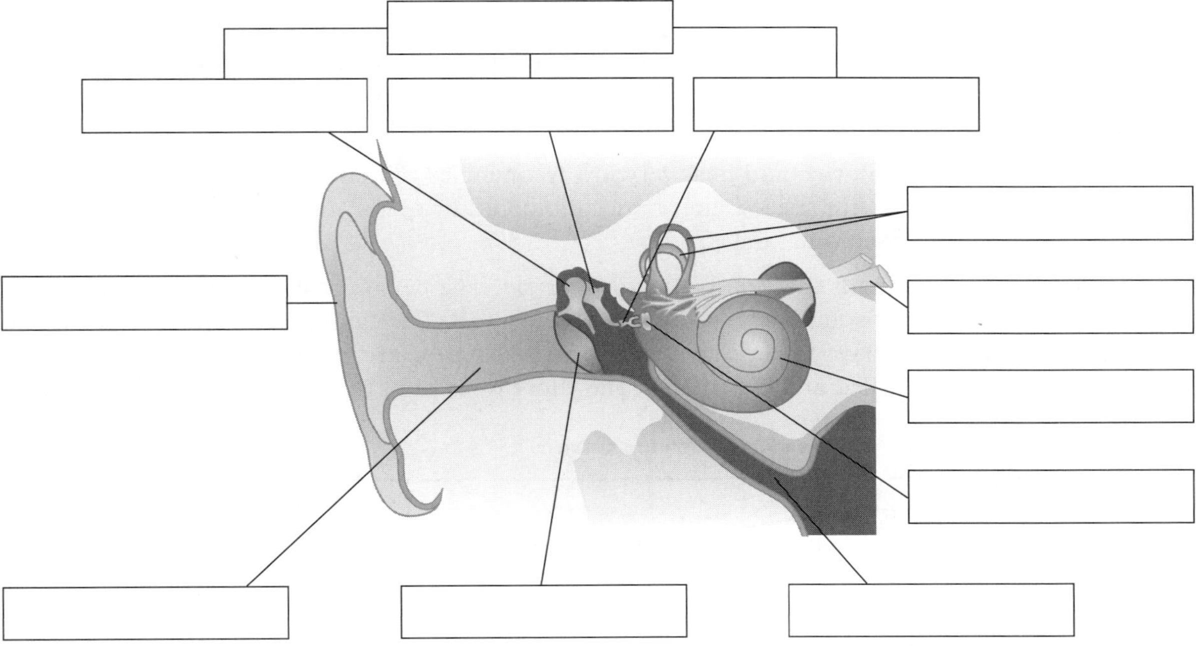

Figure 4.4.1 Structure of the ear

1 Use coloured pencils to lightly shade the outer ear yellow, the middle ear red and the inner ear blue.

2 Complete the table below by matching the bolded terms in question 1 with their correct functions in the table below. Write the correct term next to its function.

Part of the ear	Function
	Electrical impulses travel along this to the brain, which interprets them as sound.
	Vibrations pass from the middle ear into the inner ear through this part.
	This is a spiral-shaped tube filled with fluid and tiny hairs that move in response to sound vibration.
	These three tiny bones make the vibration of sound larger.
	These are filled with fluid and give us our sense of balance.
	Sound travels along the ear canal and makes this flap of tight skin vibrate.

RATE MY UNDERSTANDING Shade the face that shows your rating			

4.5 Human hearing range

Science inquiry skills

FOUNDATION | **STANDARD** | ADVANCED

The human ear can detect frequencies of sound from 20 Hertz (Hz) up to 20 000 Hertz (Hz). Hz is the symbol for Hertz, the international measure of frequency, which measures how often something happens in one second. The frequencies of sounds made when you speak are inside this hearing range, between 600 and 4800 Hz. As you age, you lose the ability to hear many higher-frequency sounds. After the age of 65, most people can only hear sounds up to 5000 Hz.

The human ear does not respond equally well to all frequencies of sound. Low-frequency sounds need to be at a much higher intensity before we can hear them, compared to sounds at the higher intensities of 2000 to 4000 hertz. This is beneficial, because if we could hear very low frequencies as well as we hear the frequencies we use for speech, the sound of blood flowing through our head would be ringing in our ears!

frequency (*n*) the number of times a sound wave is produced in a period of time

intensity (*n*) how strong or powerful something is

slogan (*n*) a short and striking or memorable phrase used in advertising

threshold (*n*) limit, border

Figure 4.5.1 shows the sounds that can be heard by an average person at different frequencies and intensities. The intensity of sound is measured in decibels (dB). We cannot hear sounds outside the threshold of the hearing curve. Sounds above the pain level curve are felt as pain rather than sound. Sounds we can hear are called audible sounds, and those we cannot hear are said to be inaudible.

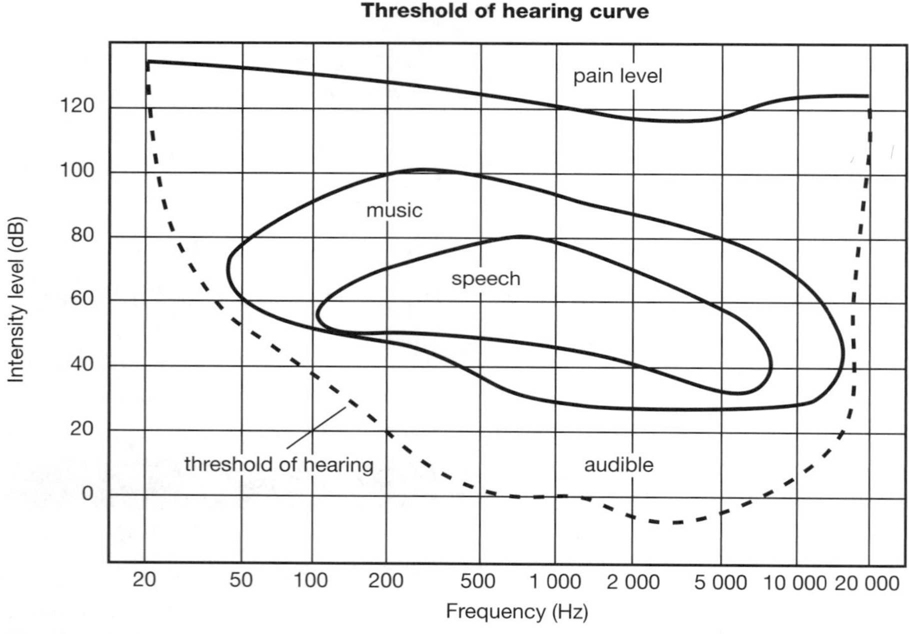

Figure 4.5.1

(1) Use a coloured pencil to shade the regions on Figure 4.5.1 that are inaudible to humans.

(2) Use the graph to predict whether each sound listed in the table is audible (heard) or inaudible (not able to be heard) by an average person.

Audible and inaudible range of human hearing		
Frequency (Hz)	**Intensity (dB)**	**Audible/Inaudible**
20	80	
10 000	40	
100	20	
15 000	60	
20 000	60	

3 The human ear is most sensitive to sounds with frequencies around 3000 Hz.

(a) Identify this point on the threshold of hearing curve in Figure 4.5.1 with an 'X'.

(b) Explain how this sensitivity is indicated by the shape of the hearing curve.

____

____

4 Thunder has a frequency of about 40 Hz. Use Figure 4.5.1 to identify the minimum sound intensity level that this needs to be, in order to be heard by humans.

________________ dB

5 A vacuum cleaner usually has a sound intensity of about 80 dB and operates at a frequency of 750 Hz. Identify this point on Figure 4.5.1 with a small 'V'.

6 The sound of a jet at take-off can be of an intensity level of 120 dB and frequency of 1700 Hz. Identify this in Figure 4.5.1 as a small diagram of an aeroplane.

7 Explain why standing on the tarmac to farewell a friend on an aircraft can be a painful experience (and not good for your hearing).

____

____

8 Exposure to loud sounds for periods of time can result in permanent hearing loss. Conduct research into hearing damage and loud sounds.

(a) What is the minimum sound intensity level known that can cause damage?

____

____

(b) Describe the symptoms suffered by a person with tinnitus.

____

(c) Imagine you own a printing business that exposes its workers to excessive noise levels. Describe three steps you could take to reduce the risk to employees.

1 ____

2 ____

3 ____

9 Figure 4.5.2 shows the image that is used to remind people to wear hearing protection. Create a slogan that could encourage people working in noisy environments to use hearing protection.

____

____

Figure 4.5.2

RATE MY UNDERSTANDING
Shade the face that shows your rating  

4.6 Reflection from plane mirrors

Science understanding

FOUNDATION | **STANDARD** | ADVANCED

1 Identify the following features in Figure 4.6.1 by:

(a) colouring the incident ray red

(b) colouring the reflected ray blue

(c) pointing to the normal with an arrow

(d) showing the angle of incidence using a star

(e) showing the angle of reflection using a cross.

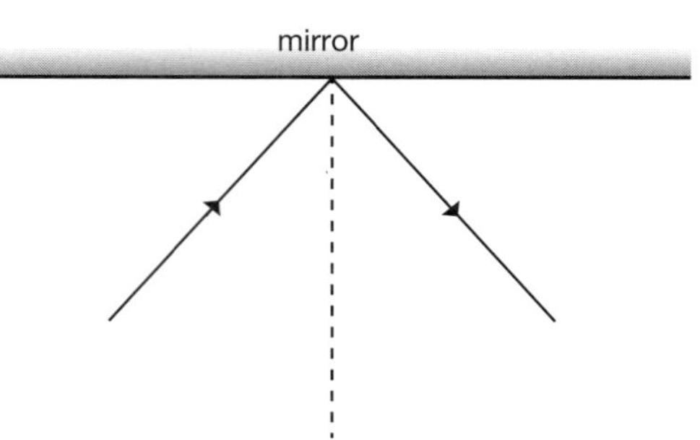

Figure 4.6.1

2 Given that there is a total of 180° making up the angles in any straight line, use your understanding of reflection to calculate the size of each angle shown in Figure 4.6.2.

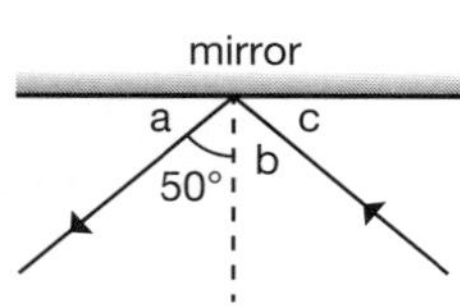

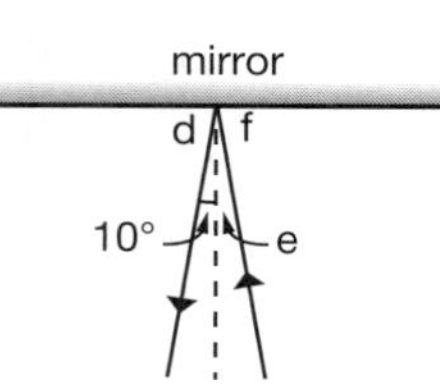

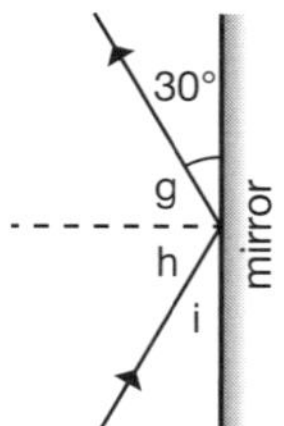

Figure 4.6.2

a = ____________ **d** = ____________ **g** = ____________

b = ____________ **e** = ____________ **h** = ____________

c = ____________ **f** = ____________ **i** = ____________

3 In Figure 4.6.3, ray 1 hits a mirror that is joined at right angles to a second mirror. Draw the path of ray 1 beyond this point.

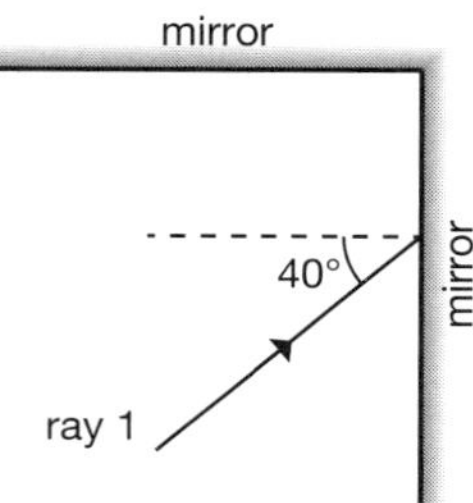

Figure 4.6.3

4 In Figure 4.6.4, Xavier stands at point X in front of a plane mirror. In the same room are an apple (A), a basket (B), a cantaloupe (C), a doughnut (D) and an egg (E). Identify which objects Xavier can see when he looks into the mirror, by drawing a circle around them.

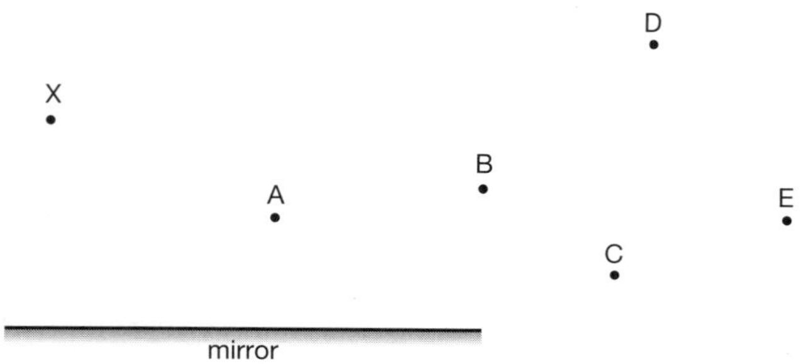

Figure 4.6.4

RATE MY UNDERSTANDING
Shade the face that shows your rating

4.7 Refraction of light

Science understanding

FOUNDATION	STANDARD	ADVANCED

When light travels from one substance into another, it bends or refracts. The light bends because it travels at a different speed in the new substance. The amount of bending, or refraction depends upon how much the speed of light changes in the new substance.

Three rays of light were passed from air into water at different angles of incidence. The angle of refraction measured for the three rays are listed in Table 4.7.1.

Table 4.7.1

Angles of incidence and refraction—from air into water		
Ray	**Angle of incidence, *i*°**	**Angle of refraction, *r*°**
1	10	7.5
2	30	22.1
3	50	35.2

1. Use a protractor to sketch the path of rays 1 to 3 as they enter water from the air. Show these paths on Figure 4.7.1 using a pencil to sketch each ray.

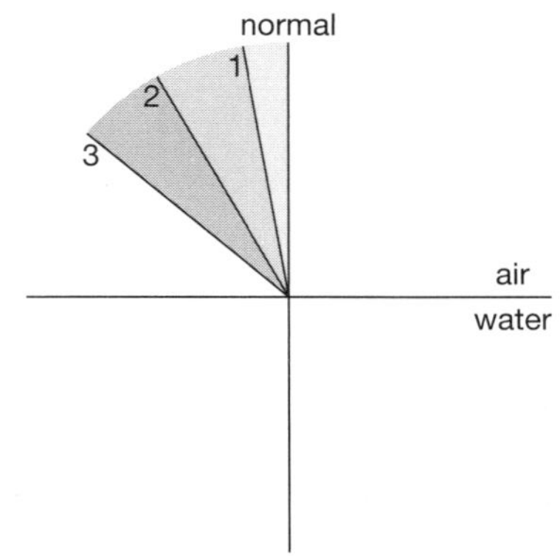

Figure 4.7.1

2. Use Table 4.7.1 data to circle the correct alternative in each statement below:
   (a) Ray 1 bends *towards/away from* the normal as it enters water from air.
   (b) Ray 2 bends *towards/away from* the normal as it enters water from air.
   (c) Ray 3 bends *towards/away from* the normal as it enters water from air.

3. If the light rays shown in Table 4.7.1 were to travel in the opposite direction, from water into air, then these rays would trace the same path as shown in Figure 4.7.1, but in the opposite direction. Use this information to complete Table 4.7.2.

Table 4.7.2

Angles of incidence and refraction—from water into air		
Ray	**Angle of incidence, *i*°**	**Angle of refraction, *r*°**
1	7.5	
2	22.1	
3	35.2	

4. Use the data in Table 4.7.2 to circle the correct alternative in each statement below:
   (a) Ray 1 bends *towards/away from* the normal as it enters air from water.
   (b) Ray 2 bends *towards/away from* the normal as it enters air from water.
   (c) Ray 3 bends *towards/away from* the normal as it enters air from water.

RATE MY UNDERSTANDING Shade the face that shows your rating	☺	😐	☹

4.8 Refraction and total internal reflection

Science understanding

FOUNDATION | STANDARD | **ADVANCED**

Light bends towards the normal when entering a material of greater refractive index. Light bends away from the normal when entering a material of lower refractive index.

1. Predict where the path of each ray of light will continue in (a), (b), (c) and (d) by drawing the light ray on each diagram in Figure 4.8.1.

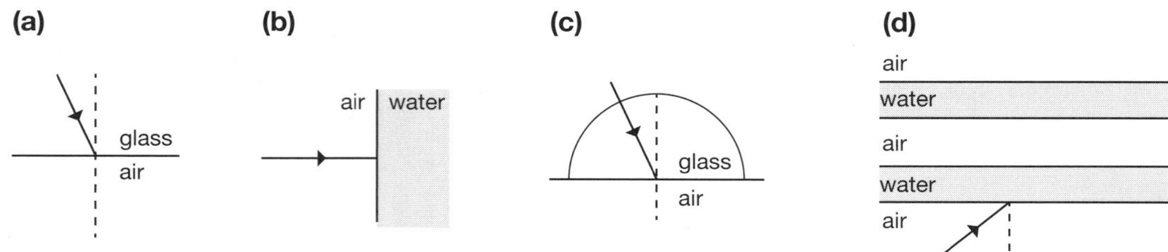

Figure 4.8.1

2. Figure 4.8.2 shows a ray of light being refracted through materials A, B and C. Analyse the diagram and list A, B and C from lowest to highest refractive index:

(lowest) ____________,

____________,

____________ (highest)

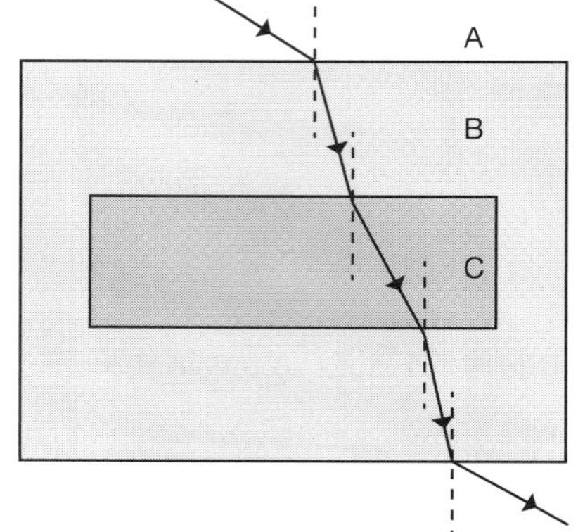

Figure 4.8.2

3. When light travels into a substance of lower refractive index, there is a critical angle at which the light is refracted at 90° to the surface. For angles larger than the critical angle, light is reflected from the surface. The critical angle for light passing from water into air is 48.8°.

Extend the rays shown in Figure 4.8.3 to show the expected path each will follow when entering air from water.

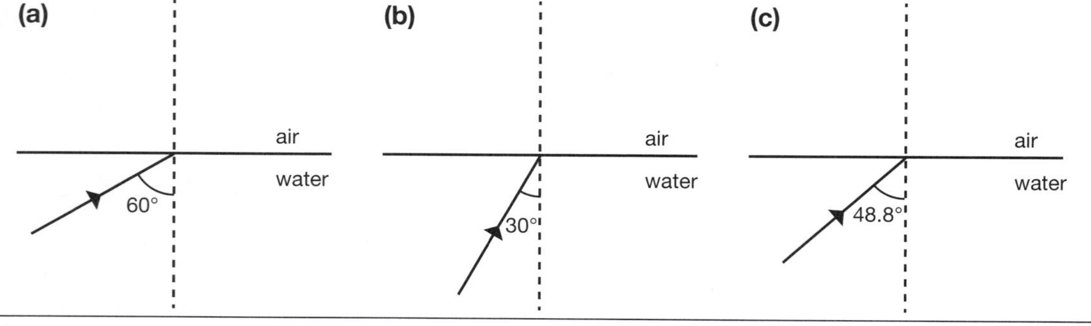

Figure 4.8.3

RATE MY UNDERSTANDING
Shade the face that shows your rating

4.9 Treating preventable blindness

Science as a human endeavour

FOUNDATION	STANDARD	ADVANCED

Four out of every five blind people living in developing countries would not be blind if they had access to basic healthcare and treatment. This means that around 80% of the world's blindness could be prevented. Almost half of these cases of blindness are due to cataracts. Cataracts are a clouding of the lens of the eye and can happen as a person ages. Trachoma can also cause blindness. Trachoma is an infectious eye disease that spreads through contact with an infected person's hands or clothes. It can lead to a condition in which the eyelids turn inwards and scrape on the cornea which can eventually lead to blindness. Diabetic retinopathy, glaucoma and age-related macular degeneration are other leading causes of blindness that could be treated or prevented through early diagnosis.

1 In pairs, research the leading causes of preventable blindness in the world today and outline treatment options. Create six questions to guide your research using the questions starters what, where, when, why, who and how. Record them on the chart on the following page.

Some key ideas that could be included into your questions and research include:

- the causes of preventable blindness, such as cataract, trachoma, diabetic retinopathy
- how these conditions could be diagnosed and treated
- the prevalence of preventable blindness amongst Indigenous Australians
- the work of the Fred Hollows Foundation
- why two-thirds of the world's blind are women.

2 Find four different websites to help you answer your questions about preventable blindness.

You should always check the usefulness, reliability and relevance of websites. The website checklist below gives criteria to help you assess websites. Write the URL at the top of each column. Place ticks or crosses against each criterion as you evaluate the website. Use the most useful websites and summarise the information you find for each of the six questions in the box on the following page.

Website Checklist	
URL 1:	URL 3:
URL 2:	URL 4:

Positives	URL 1	URL 2	URL 3	URL 4
I can identify who is responsible for the content on this site.				
The site has been updated in the last 3 to 6 months.				
This site was created by a credible person or organisation.				
The website has links to other credible websites.				
The links on the website lead you to other good information.				
The site has a .gov or .edu suffix.				
The site has useful contact information.				
The information on this site is similar to other sites I have found.				
The main purpose of this site is to provide facts (not opinions).				
Pictures on the site are helpful and have not been manipulated.				
Negatives				
The site is biased towards an opinion or point of view.				
The site contains spelling errors and broken links.				
The main purpose of the site is to sell a product.				
The site has no links to other credible websites.				

4.9 Treating preventable blindness

What	Question:
	Information:
Where	Question:
	Information:
When	Question:
	Information:
Who	Question:
	Information:
Why	Question:
	Information:
How	Question:
	Information:

RATE MY UNDERSTANDING
Shade the face that shows your rating   

4.10 Literacy review

Science understanding

FOUNDATION	STANDARD	ADVANCED

1 Recall your knowledge of heat, light and sound by completing the words below using the clues provided.

Clue	Word
Substance that allows heat to flow	c __ __ __ __ __ __ __ __
Heat transfer through a liquid or a gas	__ __ __ v __ __ __ __ __ __
Heat transfer that can occur through a vacuum	__ __ __ __ __ t __ __ __
Instrument used to measure temperature	__ __ __ __ m __ __ __ __ __ __
Energy that enables us to see	l __ __ __ t
This energy is caused by vibrations	__ __ __ nd
A region of high pressure in a sound wave	__ __ __ p __ __ ss __ __ __
A region of low pressure in a sound wave	__ __ __ __ f __ c __ __ __ __
Waves at the beach are of this type	tr __ __ __ __ __ __ __ __
Sound waves are of this type	__ __ __ git __ __ __ __ __ __
A high-pitched sound has a high...	__ __ __ qu __ __ __ __
The frequency of a sound is measured in	__ __ __ __ z
The loudness of a sound is measured in	__ __ __ __ b __ __ __
Light is a form of energy called ____________ radiation	__ __ __ c __ r __ ma __ __ __ __ __ __
An object that releases or emits light is...	__ u __ i __ o __ __
What light does when it hits a mirror	r __ __ __ __ __ __ s
Rays of light do not cross when producing this type of image	__ __ r __ u __ l
Rays of light do cross when forming this type of image	r __ __ l
Bending of light	__ e __ __ __ c __ __ __ __
A measure of how easily light passes through a material is called its ____________ index	r __ __ r __ __ __ __ __ __
At greater angles of incidence than this, light is totally internally reflected	c __ __ __ __ c __ __
A transparent piece of plastic or glass that is shaped to curve outwards or inwards	__ __ __ s
Lens that bulges outwards	c __ __ __ __ x
Lens that curves inwards	c __ __ c __ __ __
The part of the eye on which an image forms	__ __ t __ __ __

RATE MY UNDERSTANDING
Shade the face that shows your rating

4.11 Thinking about my learning

1 Reflect on what you have learned about heat, light and sound by completing each section of the learning wheel below:

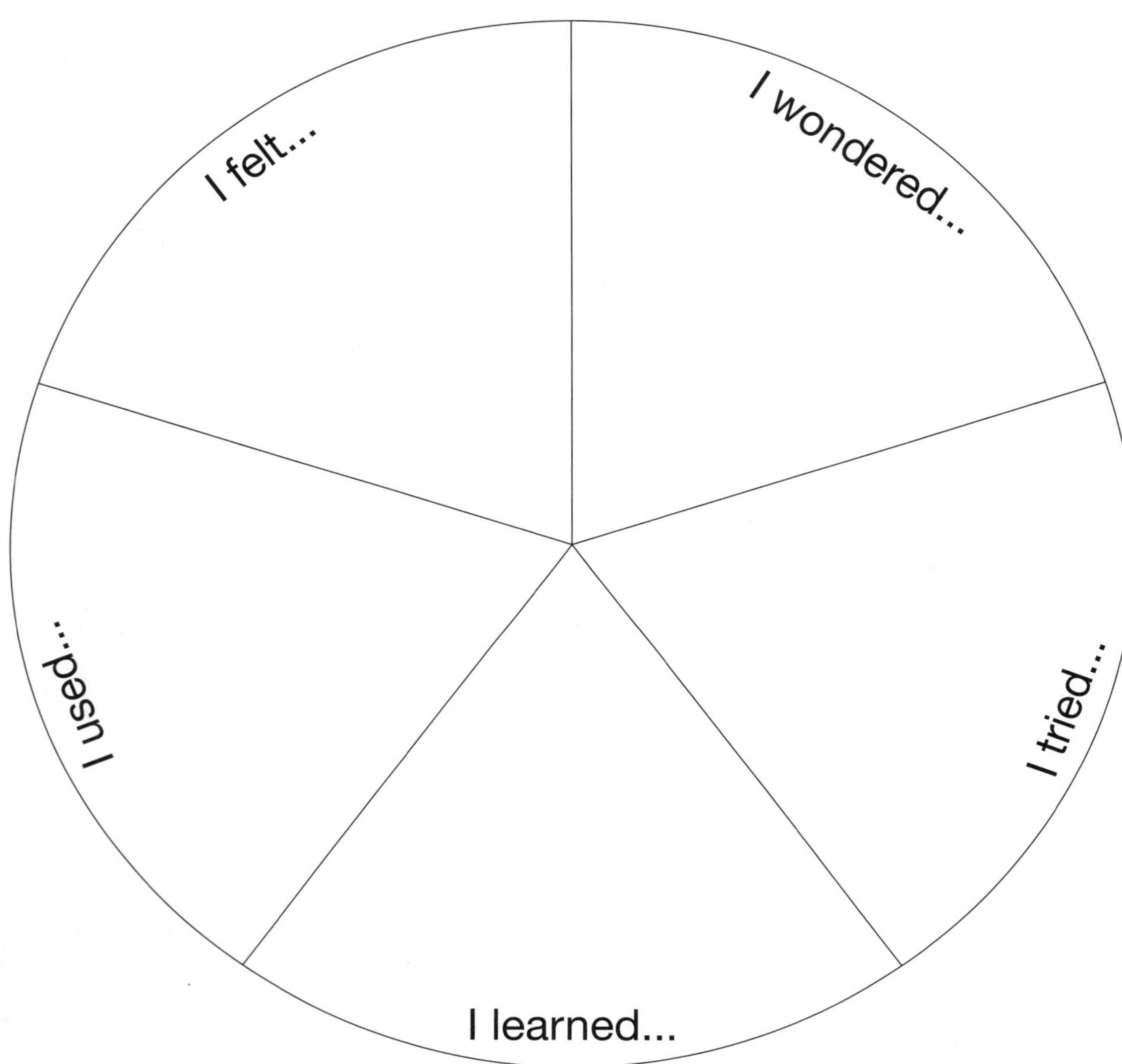

CHAPTER 5

Electromagnetic radiation

5.1 Knowledge preview

Science understanding

FOUNDATION | **STANDARD** | ADVANCED

1 Six different types of waves and their uses are shown below. These waves make up most of the electromagnetic spectrum. In the space below each image, write what you know about each wave type.

radio	image using ultraviolet	microwave
image using infrared	gamma	visible light

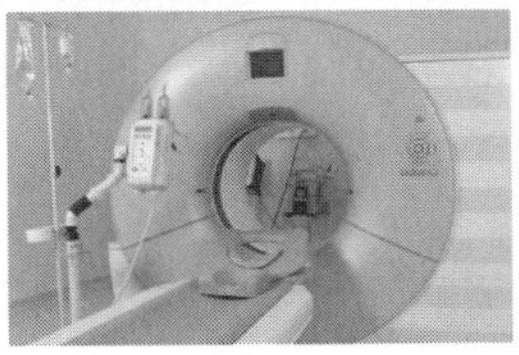

2 **(a)** The wave that produced the image below is not included in the examples of electromagnetic waves above. What is the name of the wave that produced this image?

(b) Explain what this type of wave is used for.

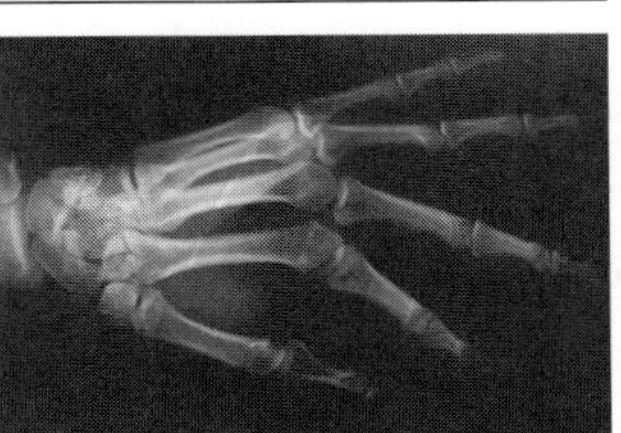

5.2 The wave equation

Science inquiry skills

FOUNDATION | **STANDARD** | ADVANCED

Processing & Analysing 

The speed, wavelength and frequency of any wave depend upon each other and are linked by a formula called the wave equation:

$v = f\lambda$

where

v = speed of wave (m/s)

f = frequency of wave (Hz)

λ = wavelength of wave (m)

The equation can be rearranged to calculate:

frequency as: $f = v/\lambda$

wavelength as: $\lambda = v/f$

Worked example

Su plays a note on her violin of frequency 4100 Hz. Given that the speed of sound in air is 330 m/s, calculate the wavelength of this sound wave.

$\lambda = v/f$

$= 330/4100$

$= 0.08$ m

This means that the sound wave has a wavelength of approximately 8 cm.

1. Use the speed of sound in air as 330 m/s to answer the following questions. Round your answers to two decimal places.

(a) On a piano, Carla plays middle C, which has a frequency of 256 Hz. Calculate the wavelength of the sound wave producing this note.

____

____

(b) Sebastian is annoyed by a mosquito buzzing near his ear. The buzzing sound is made by the wings of the mosquito, which beat about 600 times per second. Calculate the wavelength of the sound wave produced.

____

____

(c) An elephant makes a very low sound of 20 Hz as it communicates to its herd. Calculate the wavelength of this sound.

____

____

(d) A mobile phone makes sound of frequency 16 000 Hz when an SMS is received. Calculate the wavelength of this sound.

____

____

5.2 The wave equation

(2) Use the speed of electromagnetic radiation as 300 000 000 m/s to answer the following questions. Round your answers to two decimal places and express them using scientific notation. For example 300 000 000 m/s = 3×10^8 m/s. Be sure to use standard units of m, m/s and Hz in your calculations.

(a) Microwave radiation used to cook food in a particular oven has a wavelength of 3 cm. Convert this wavelength into metres, then use this value to calculate the frequency of the radiation.

____

____

____

(b) Microwaves of wavelength 30 cm are used to transmit signals in a mobile phone network. Calculate the frequency of this radiation.

____

____

____

(c) A UHF TV signal uses a radio wave of 80 cm. Calculate its frequency.

____

____

____

(d) A 500 kHz radio wave is used to transmit an AM radio signal. Calculate the wavelength of these radio waves.

____

____

____

(e) A truck driver uses a 20 MHz (20 000 000 Hz) shortwave radio for communications. What is the wavelength of the radio waves being used?

____

____

____

RATE MY UNDERSTANDING
Shade the face that shows your rating

5.3 Butterflies and mobiles

Science as a human endeavour

FOUNDATION	STANDARD	ADVANCED

morpho butterfly (*n*) a butterfly found in South America

spectrum (*n*) the range of visible colours

substrate (*n*) layer beneath

A company called Qualcomm has copied how a butterfly wing produces colour to develop a new type of display for mobile phones and other devices. Rather than using a battery-powered light source for the phone display, it uses light from the room or sunlight as its source. This means that the phone's battery lasts longer between charging. The phone contains tiny reflective units that cause light waves entering it to overlap, or interfere with other light waves. Some light waves cancel each other out and some add to each other. This produces different colours in the same way that the wings of a Morpho butterfly cause light to interfere and create colours.

The basic unit of the display that causes the interference of incoming light is a tiny square of glass and metal (called an interferometric modulator display element or IMOD). This unit has two layers of reflective surfaces with an air gap between them (Figure 5.3.1). The air gap can be closed or opened by applying a small voltage across it. When it opens, light waves can reflect off the bottom-reflecting membrane and the top-reflecting surface (the 'thin film stack'). If the gap is the right size, then the two reflected waves of light add to each other and appear a certain colour, such as blue. When the air gap closes or is collapsed, no light reflects off the surfaces and they appear black.

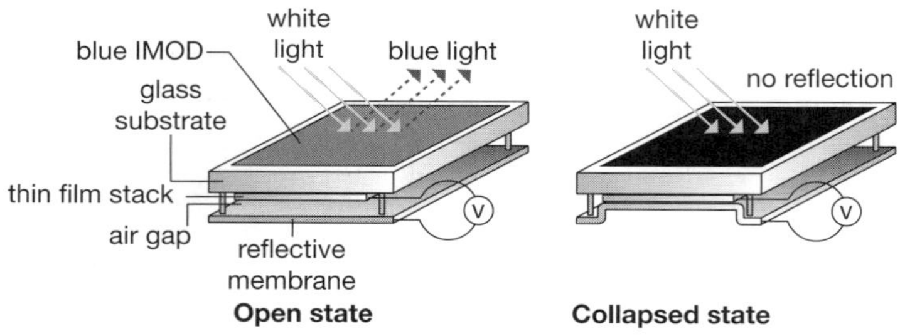

Figure 5.3.1 IMOD with open and collapsed membranes

To make an IMOD element produce light of a particular colour, the air gap is manufactured in three different sizes. Red is the largest gap and blue is the smallest. Green is a medium-sized gap. You can see some sets of the basic units in Figure 5.3.2.

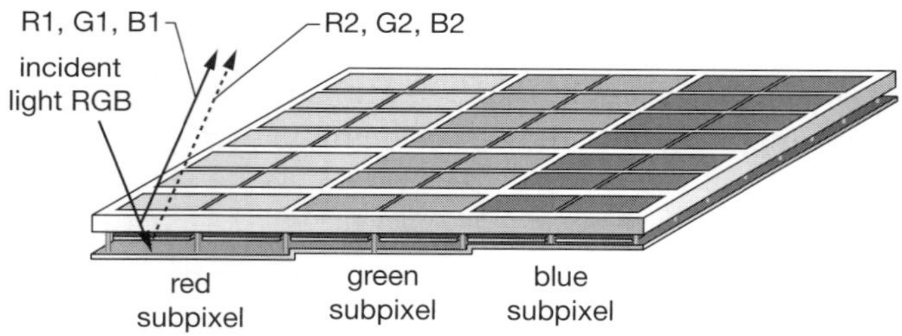

Figure 5.3.2 IMOD membrane positions to make red, green and blue colours

5.3 Butterflies and mobiles

When the screen needs to produce particular colours, the three different colours of IMOD elements are opened or closed in sets of three to create a colour effect. You can see an example in Figure 5.3.3. To produce a yellow colour, you would need to allow red and green light to mix, but not let any blue light through. So the blue IMOD element is turned off, and red and green light reflect out of the two nearby elements. Your eye detects this reflected light as yellow. By turning different combinations of elements off or on, the screen can produce all the colours of the spectrum.

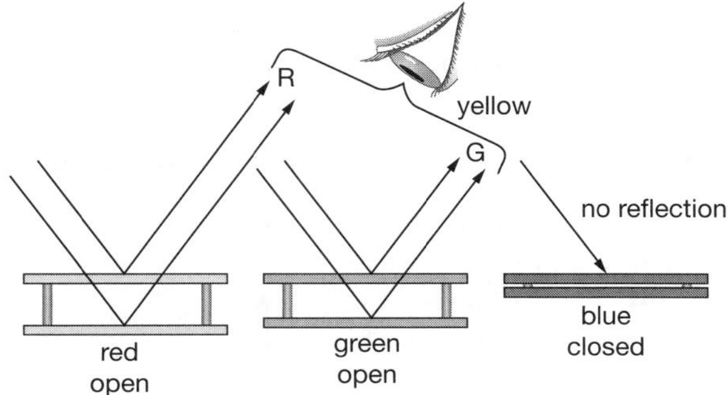

Figure 5.3.3 Mixing IMOD colours to make other colours

1. What is the main advantage of a mobile phone display that uses reflected light?

2. Name the animal that was the inspiration for this idea.

3. Name the basic unit of the display.

4. Explain how the structure of an IMOD element affects the colour of light it produces.

5. Explain how a red, a green and a blue IMOD elements next to each other can produce yellow light.

6. What would have to happen to the three IMOD elements to produce the following colours?

(a) blue light

(b) purple light

HINT
Red and blue make purple

RATE MY UNDERSTANDING
Shade the face that shows your rating  

5.4 Sunspot activity

Science understanding

FOUNDATION | STANDARD | **ADVANCED**

Sunspots are large, dark areas on the surface of the Sun. Sunspots indicate the presence of strong magnetic fields. Intense magnetic activity can cause violent eruptions called solar flares that release large amounts of electromagnetic energy including gamma rays and X-rays. When there are more sunspots on the Sun, there is more magnetic activity. The table in Figure 5.4.1 shows the sunspot number recorded from 1995 to 2014. The graph in Figure 5.4.1 shows the sunspot numbers recorded each year from 1920 to 1994. The sunspot number is an average calculation of the number of sunspot groups and individual sunspots that appeared on the Sun during a particular year.

Number of sunspots 1995 to 2014

Year	Sunspot number
1995	17.9
1996	8.6
1997	21.5
1998	64.3
1999	93.3
2000	119.0
2001	110.9
2002	104.0
2003	63.7
2004	40.4
2005	29.8
2006	15.2
2007	7.5
2008	2.7
2009	8.7
2010	9.9
2011	55.7
2012	57.6
2013	64.7
2014	79.3

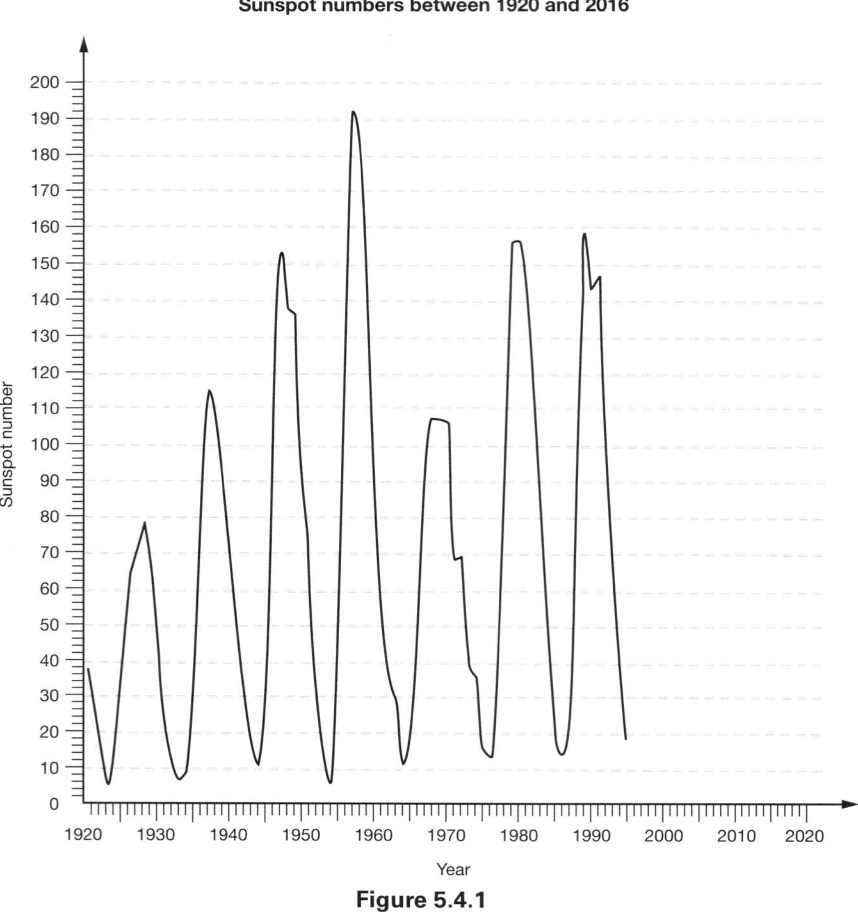

Figure 5.4.1

1. Complete the graph by plotting the sunspot numbers for 1995 to 2014.

2. The variation in sunspot activity is periodic. Being periodic means that sunspot activity repeats itself, forming a cycle. This is called the sunspot cycle. Identify years of maximum solar activity with 'Max' on your graph, and those with of minimum activity with 'Min'.

3. Use your graph to calculate the length of the sunspot cycle, or the time between successive maximum or minimum solar activity. ______________ years

4. **(a)** State the number of sunspots in the year you were born. ______________

   (b) Use the trend of the graph to predict the sunspot number in the year you will turn 21.

   ____

RATE MY UNDERSTANDING
Shade the face that shows your rating

5.5 Night vision

Science as a human endeavour

FOUNDATION | STANDARD | **ADVANCED**

Scientists reveal how snakes 'see' at night

by Marlowe Hood, 15 March 2010 AFP

Scientists revealed for the first time how some snakes can detect the faint body heat exuded by a mouse a metre away with enough precision and speed to hunt in the dark.

It has been known for decades that rattlesnakes, boas and pythons have so-called pit organs between the eye and the nostril that can sense even tiny amounts of infrared radiation—heat—in their surroundings.

Even with tiny patches covering its eyes, the snake has shown the ability to track and kill prey blindfolded.

'In this case, the infrared radiation is actually detected inside the pit organ as heat,' Molecular biologist David Julius at the University of California, San Francisco, said in a phone interview. 'We found the molecule responsible.'

A very thin membrane inside the pit organ—essentially a hollow, bony cavity—warms up as the radiation enters through an opening in the skin, he explained.

Because the membrane is in a hollow space, it is exquisitely sensitive to changes in temperature.

'The heated tissue then imparts a signal to nerve fibres to activate the receptors we have identified,' known as TRPA1 channels.

The neurochemical pathway involved suggests that snakes feel heat rather than see it.

'It is amazing to think that random mutation could have come up with the same kind of solution more than once,' Julius said.

1 List three types of snakes that can detect infrared radiation. ______________________

____

2 Describe the structure of the pit organ of a snake. ______________________

____

____

(3) Identify stages in the detection of infrared radiation in the pit organ of a snake by completing the following.

____________ ____________ enters the pit organ.	→	This warms up a very thin ____________ which	→	transmits a signal to ____________ ____________	→	receptors called ____________ channels.

(4) Explain how a snake would use its ability to sense heat to its advantage. ______________________

____

____

[5] Discuss three ways in which your life would be different if you had a heat-sensing pit organ between your nostrils and your eyes.

____

____

RATE MY UNDERSTANDING
Shade the face that shows your rating

5.6 The discovery of X-rays

Science as a human endeavour

FOUNDATION	**STANDARD**	ADVANCED

At times something happens in an experiment that was not expected. Some of the most remarkable scientific discoveries have been made by accident. To make a breakthrough, a scientist needs to make careful observations and to think 'outside the square'. Wilhelm Roentgen did just that when he discovered X-rays.

> **'outside the square'** (*adj*) to think creatively; to change the way you think

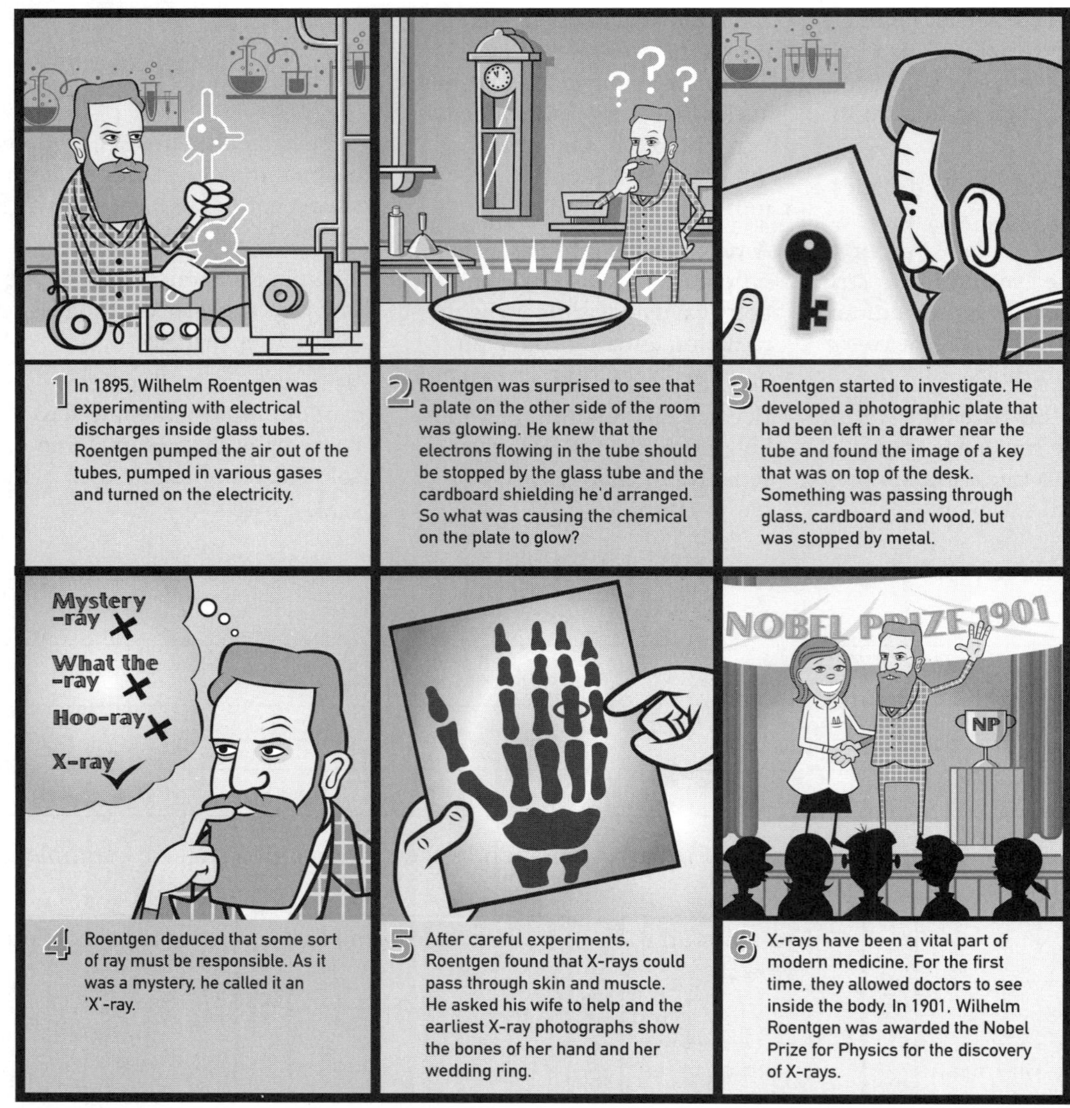

1 In what year did Wilhelm Roentgen discover X-rays? ______________________

2 Describe what Roentgen was initially investigating with his gas tubes.

____

____

3 Roentgen covered the glass discharge tube with cardboard. Describe what he saw happening on the other side of the room that made him curious.

____

____

4 How did Roentgen realise that something was able to pass through glass, cardboard and wood but was stopped by metal in the room in which he was working?

____

____

____

5 If the radiation Roentgen produced had been able to pass through metal in addition to glass, cardboard and wood, predict how the photographic plate would have appeared.

____

____

6 Explain why Roentgen called the radiation X-rays.

____

____

7 Looking at the diagram of Roentgen's wife's hand, list the properties of X-rays that make them a very useful tool in medical diagnosis.

____

____

8 What official recognition did Roentgen receive for his discovery?

____

RATE MY UNDERSTANDING
Shade the face that shows your rating

5.7 Creating a false-colour X-ray image

Science understanding

FOUNDATION | **STANDARD** | ADVANCED

When you take a photo of a friend, you are capturing an image produced by visible light. Images can also be created by capturing wavelengths of electromagnetic radiation that are invisible to the human eye, such as radio waves, microwaves or X-rays. These images are said to be 'false colour', because different colours correspond to a specific range of intensities of radiation.

The galaxy cluster Hydra A is located 840 million light years from Earth. It consists of hundreds of galaxies and huge clouds of extremely hot gas which are all held together by gravity. Matter in Hydra A is at temperatures up to 100 million degrees Celsius. X-rays are produced at these extreme temperatures. The image below shows what Hydra A looks like in visible light. Observatories such as the Chandra X-ray Observatory can detect the presence of high-energy X-rays in this cluster.

0	0	1	1	1	1	2	2	2	2	2	2	1	1	0
0	0	1	1	2	3	2	2	2	2	2	2	2	2	1
0	1	1	1	2	3	3	3	3	2	3	2	2	1	1
1	1	1	2	3	3	3	4	3	3	3	3	3	2	1
1	1	2	2	3	4	4	5	4	3	3	3	3	2	1
1	1	2	3	3	4	5	5	5	4	5	3	2	2	2
1	1	2	3	3	4	6	6	5	5	5	3	2	2	1
1	1	2	3	4	5	5	6	5	4	4	3	2	2	1
1	1	2	3	4	5	5	4	4	4	3	3	3	2	1
1	1	2	3	3	4	4	4	4	4	4	3	2	2	1
1	1	2	2	3	3	4	4	3	3	3	3	2	2	1
1	1	2	2	2	3	3	3	3	2	2	2	2	2	1
1	1	2	2	2	3	3	3	3	2	2	2	2	2	1
0	1	1	1	2	2	2	2	2	2	2	1	2	2	1
0	1	1	1	1	1	2	2	2	1	1	1	1	1	0

1. Using the grid above, create your own false-colour image that shows the intensities of the X-ray emissions found in the cluster. The numbers shown represent very intense X-ray emission (6) to no X-ray emissions (0). To display these variations, use the colour code: 6 = white, 5 = yellow, 4 = orange, 3 = pink, 2 = purple, 1 = blue, 0 = black.

2. Explain why it is useful to view a region of space or a target on Earth using forms of radiation other than visible light.

RATE MY UNDERSTANDING
Shade the face that shows your rating

5.8 Radiation dose

Science understanding

FOUNDATION	**STANDARD**	ADVANCED

X-rays are used in diagnostic procedures to give information about a person's body. When an X-ray or CT (computed tomography) image is taken, the patient is exposed to X-ray radiation. This radiation can damage cells in the body, and large doses of radiation may be harmful. The amount of radiation a patient experiences during diagnostic imaging has a low risk. However, it is important for people who work with X-ray radiation to monitor how much they absorb using a personal radiation monitoring device (PMD).

Every year, your body absorbs a certain amount of radiation from the background sources around you. Radiation occurs naturally on Earth in soil, food, water, the atmosphere and in vegetation. Cosmic rays from space also contribute to background radiation. There are greater numbers of these rays at higher altitudes. Some forms of radiation are more damaging than others. A quantity called the dose equivalent is a measure of the effect of the radiation that is absorbed. It is measured in a unit called the sievert (Sv). The average annual background radiation dose that is absorbed by each person in Australia is about 2.5 mSv.

altitude (*n*) the height above sea level

CT (*n*) computed tomography – a medical imaging method

diagnostic procedure (*n*) a medical test to find out what is wrong with someone

mammogram (*n*) breast scan

medical procedure (*n*) anything medically done to improve the health of a patient

The Background Equivalent Radiation Time unit (the BERT) is a useful way for people to compare the radiation dose from a medical procedure to how many hours, days or weeks it would take them to absorb a similar amount of background radiation. For example, one flight from Europe to Australia exposes a person to an effective radiation dose of around 0.11 mSv. This is similar to absorbing about 15 days of natural background radiation. Table 5.8.1 compares BERT values for some medical procedures. These are included as a guide; actual measurements vary with differences between patients.

Table 5.8.1 BERT values for some medical procedures

Procedure	Approximate effective radiation dose (mSv)	Natural background radiation this is equivalent to (BERT)
chest X-ray	0.025	3 days
dental X-ray	0.06	1 week
X-ray: lumbar spine	0.47	10 weeks
X-ray: abdomen or pelvis	0.35	7 weeks
CT: head	1.6	8 months
CT: chest	3.8	1.5 years
CT: abdomen/pelvis	10.0	3 years
bone scan	4.2	1.8 years
mammogram	0.44	9 weeks

1 Describe how people working in medical imaging monitor their absorption so they are not exposed to dangerous levels of radiation.

____

____

2 List three sources of background radiation that originate from Earth.

3 What is the source of background radiation from space?

(4) Explain why aircraft crew must monitor their exposure to cosmic radiation.

5 **(a)** State the unit of the dose equivalent.

(b) What is the average amount of background radiation absorbed by a person living in Australia per year?

6 **(a)** State what BERT stands for.

(b) Explain why the BERT is used.

(7) Use Table 5.8.1 to answer the following.

(a) How long would it normally take to absorb the amount of background radiation that is absorbed in a dental X-ray?

(b) Calculate how many times larger the natural background radiation of a CT scan of the pelvis is compared to a CT scan of the head.

(c) Compare the effective radiation dose of a mammogram with that of a CT scan of the head.

RATE MY UNDERSTANDING
Shade the face that shows your rating

5.9 Literacy review

Science understanding

FOUNDATION | STANDARD | ADVANCED

1 Match the jumbled list of terms in the left column with their correct definitions in the right column. Match them by writing the correct term in the middle column next to its correct definition.

Jumbled terms	Correct terms	Definitions
dispersion		This is the term for the entire range of frequencies of electromagnetic radiation.
gamma		This type of electromagnetic wave is commonly used in cooking and in mobile phone transmission.
microwave		These rays are the highest energy form of electromagnetic radiation.
optical fibre		This band of electromagnetic waves has the longest wavelength.
radio waves		Your eyes detect this electromagnetic radiation as this sort of light.
spectrum		Your skin and eyes need to be protected from this type of radiation.
ultraviolet		This high energy radiation has been used for over a century to produce images of inside the body.
visible light		This is narrow tube made of glass or plastic that is used to transmit pulses of light.
x-rays		The splitting of white light into its component parts.

RATE MY UNDERSTANDING
Shade the face that shows your rating

5.10 Thinking about my learning

Fill in the boxes using the three Rs (remember, reveal and revise) to check your learning about electromagnetic radiation.

REMEMBER: Write or draw three things you can remember about electromagnetic radiation.

REVEAL: Write or draw two things you found really interesting about electromagnetic radiation.

REVISE: Write or draw one thing you are still unsure about and will need to recap before moving on.

CHAPTER 6

Electricity

6.1 Knowledge preview

Science understanding

FOUNDATION	STANDARD	ADVANCED

Read each question/statement in question 1 and highlight the correct answer from the four options provided.

1 **(a)** Atoms are made up of sub-atomic particles called protons, neutrons and electrons. Which statement is correct?

A protons are positively charged, neutrons negative and electrons neutral

B protons are negatively charged, neutrons positive and electrons neutral

C protons are positively charged, neutrons neutral and electrons negative

D none of the above

(b) The simple diagram (Figure 6.1.1) represents the structure of an atom.

A X indicates protons orbiting the electrons in Y

B Y indicates the nucleus

C Y indicates a cluster of neutrons

D X indicates neutrons orbiting the nucleus

Figure 6.1.1

(c) A solar cell is responsible for converting:

A heat energy into electrical energy

B light energy into electrical energy

C electrical energy into heat energy

D light energy into heat energy

(d) When using electrical appliances it is safe to:

A pull a plug out by its cord or plug

B turn off power points before changing light globes

C touch exposed electrical wires

D use a knife to remove a piece of toast while toaster is working

2 The list on the left shows a source of electric energy. The list on the right indicates the energy conversions that happen when electricity is produced. Match up the electric energy source with the correct energy conversion by drawing arrows between the two lists.

coal power
wind power
solar power
nuclear power
hydro power

light energy------electrical energy
water movement------electrical energy
chemical energy------heat energy------electrical energy
air movement------electrical energy
atomic energy------heat energy------electrical energy

6.2 Reading meters

Science inquiry skills

FOUNDATION | **STANDARD** | ADVANCED

Processing & Analysing

Many ammeters and voltmeters have two or more inputs. The reading that you take depends on what input is being used. The diagrams below show an ammeter that reads in amps (A) or milliamps (mA), depending on which terminal is being used to connect the ammeter to the circuit. For example, if the wires are connected to the 500 mA terminal, you need to read off the 500 mA scale.

1. Use this information to determine the voltage reading of each voltmeter shown below.

(a)

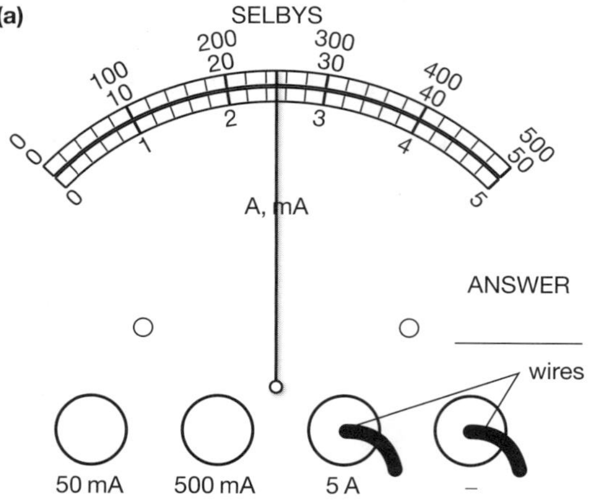

ANSWER ________

(b)

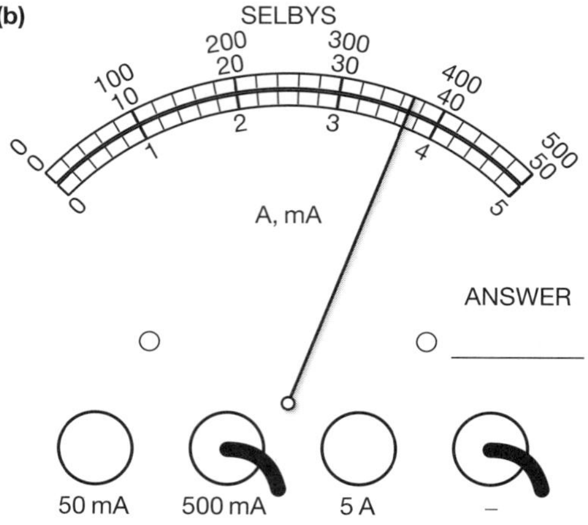

ANSWER ________

(c)

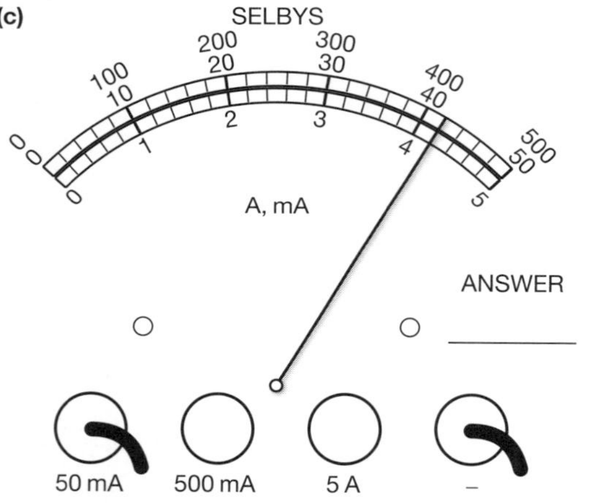

ANSWER ________

(d)

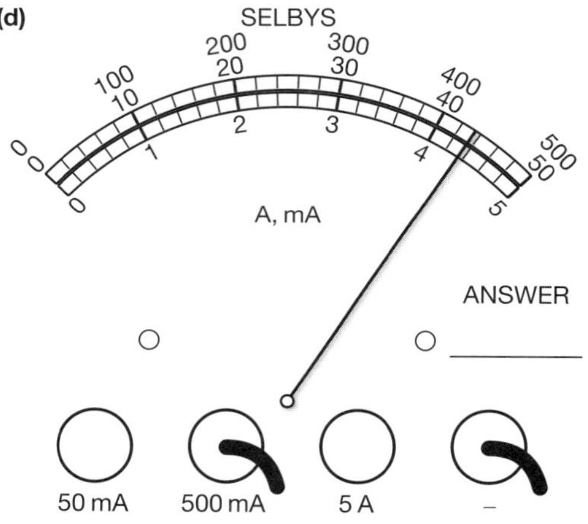

ANSWER ________

(e)

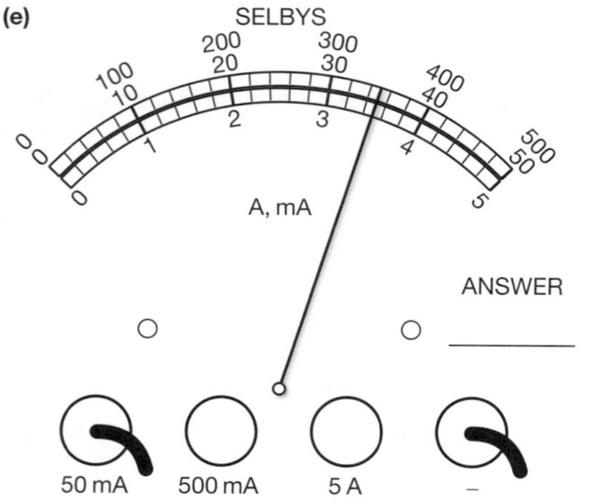

ANSWER ________

(f)

ANSWER ________

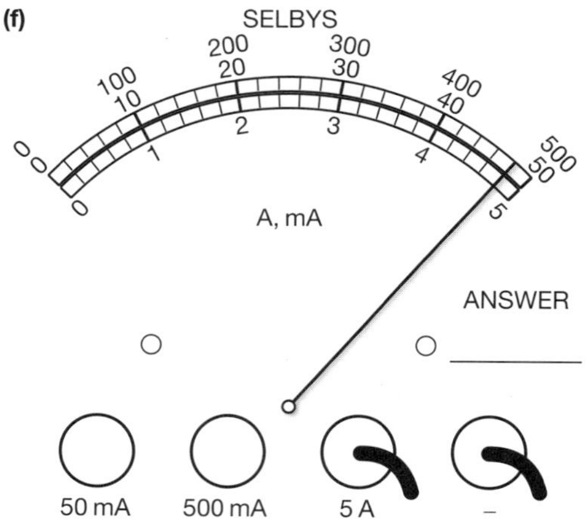

RATE MY UNDERSTANDING
Shade the face that shows your rating

6.3 Electrical circuits

Science inquiry skills

FOUNDATION | **STANDARD** | ADVANCED

Processing & Analysing

Scientists use models to help them understand what is going on. An analogy is a model that compares something that is difficult to understand with something that is easy to understand. For example, the structure of the atom is sometimes compared with the structure of the solar system. In this analogy, the Sun represents the nucleus of the atom, and the planets (which are tiny compared with the Sun) represent the electrons.

In a similar way, a simple electric circuit can be compared with the 'water circuit' that runs a decorative water wheel in a garden pond. Figure 6.3.1 compares the two.

analogy (*n*) to say that two things are alike e.g. the heart is like a pump

defined (*adj*) exact, known

represent (*v*) to stand in place of

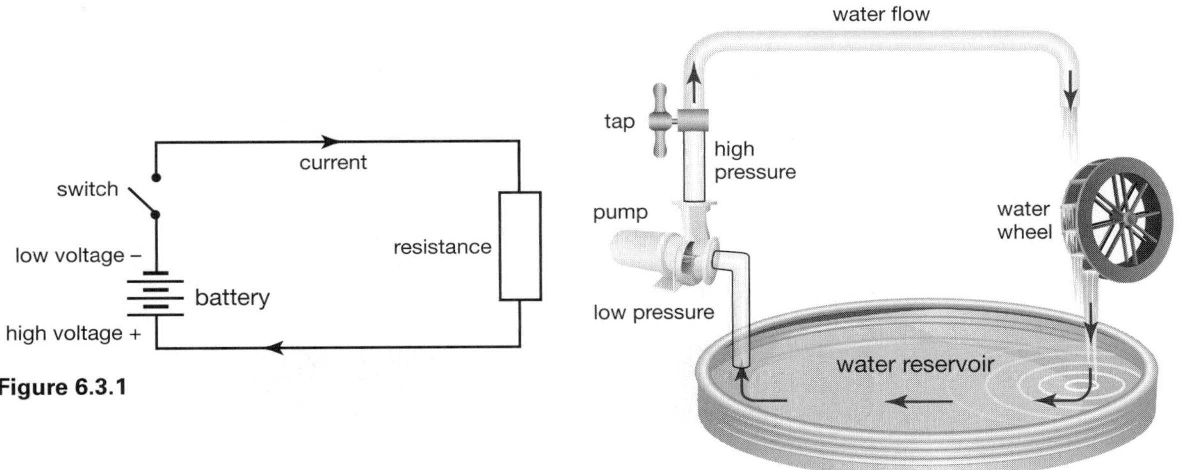

Figure 6.3.1

The two circuits are similar.

- Both have particles flowing around them. Electrons flow around the electric circuit while molecules of water flow around the water circuit.
- Both have wires or pipes connecting all their components.
- Both have a switch or tap to turn the flow on or off.
- Both need an energy source to operate. The electric circuit needs a battery while the water needs a pump.
- Both have an energy user. The electric circuit uses a light globe, while most of the energy of the water is 'used' in the water wheel itself.

Analogies are not always completely accurate descriptions of what is going on. For example, electrons do not travel in defined orbits like the planets around the Sun. Likewise the water circuit doesn't describe everything that happens in an electric circuit. When you cut a pipe, water will flow out of it. However, when you cut a wire in an electric circuit, the current stops.

6.3 Electrical circuits

1 **(a)** Describe the analogy used to explain the structure of the atom.

(b) Identify features that make it a good analogy.

(c) Identify the features that make it a poor analogy.

2 Refer to the electric circuit and water circuit analogy in Figure 6.3.1. Match each of the following components in an electric circuit with the feature in a water circuit that it is most similar to, by connecting them with lines.

Component of electric circuit
wires
switch
battery
globe

Feature in a water circuit
pump
water wheel
pipes
tap

3 **(a)** Describe ways in which the 'water circuit' analogy accurately explains what happens in an electric circuit.

(b) Describe where the analogy fails.

RATE MY UNDERSTANDING
Shade the face that shows your rating

6.4 Ohm's law

Science understanding

FOUNDATION	STANDARD	ADVANCED

A close relationship exists between the voltage, current and resistance of a circuit. If you increase the voltage supplied to a circuit, then the current through it will increase. If you increase the resistance of a circuit, then the current through it will drop. For components that obey Ohm's law, a line graph of voltage plotted against current produces a straight line. This is shown in Figure 6.4.1.

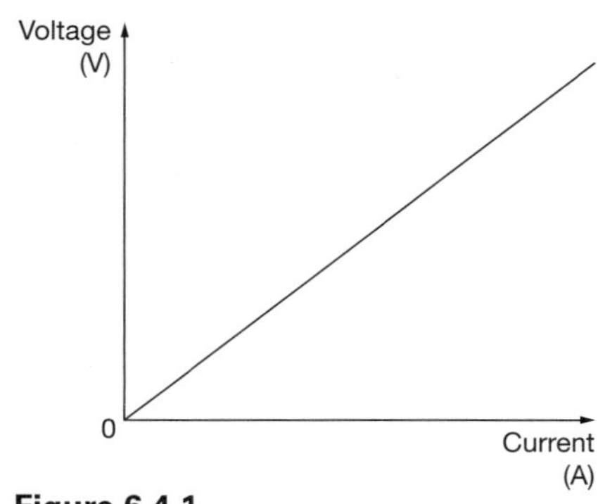

Figure 6.4.1

This close relationship can be summarised by Ohm's law, which states:

voltage = current × resistance

or

$V = IR$

where

V is the voltage (measured in volts, V)

I is the current (measured in amperes, A)

R is the resistance (measured in ohms, Ω).

1 Use words from the box below to complete the following statements.

increases	decreases	stays the same

(a) For a constant voltage, current ______________________ as resistance increases.

(b) For a constant resistance, voltage ______________________ as current increases.

(c) For a constant current, voltage ______________________ as resistance increases.

2 The graphs below were plotted for four different circuit components. Identify which graph represents a component that obeys Ohm's law.

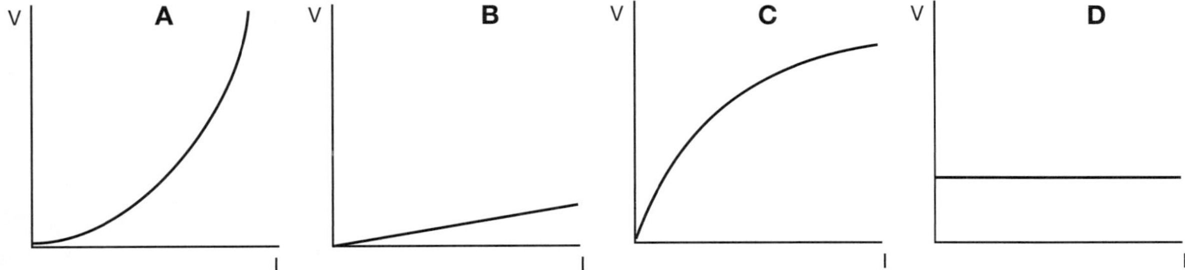

6.4 Ohm's law

3 Calculate the voltage across the following resistors.

(a) A current of 4 A is flowing through a 5 Ω resistor

(b) A current of 1.5 A is flowing through a 4 Ω resistor.

4 Calculate the current flowing through a:

(a) 3 Ω resistor when a voltage of 6 V is applied across it

(b) 10 Ω resistor when a voltage of 12 V is applied across it.

5 Calculate the resistance of a wire if it carries:

(a) 2 A when a voltage of 6 V is applied across it

(b) 1.5 A when a voltage of 12 V is applied across it

(c) 50 mA (0.05 A) when a voltage of 6 V is applied across it.

RATE MY UNDERSTANDING
Shade the face that shows your rating

6.5 Plotting Ohm's law

Science understanding

FOUNDATION | STANDARD | **ADVANCED**

A team of Year 9 students measured the current through a light globe as the voltage applied to it increased. Their results are shown in the table below.

Current though lightbulb with increasing voltage							
Voltage (V)	0	1.5	3.0	4.5	6.0	9.0	12.0
Current (A)	0	3.0	4.0	4.5	4.8	6.0	8.0

1. Plot the students' results on the graph below. Be sure to add a title for the graph.

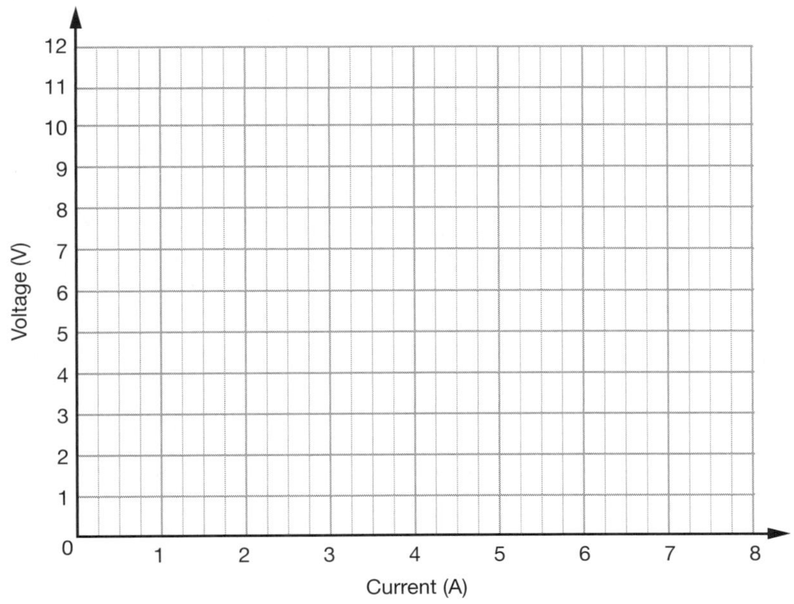

2. Use values from the above table or your graph to calculate the resistance (measured in ohms) at the following voltages.

(a) 1.5 V ______

(b) 4.5 V ______

(c) 9.0 V ______

3. **(a)** Assess whether the light globe is obeying Ohm's law or not.

______

(b) Justify your answer.

______

______

______

______

RATE MY UNDERSTANDING
Shade the face that shows your rating

6.6 Predicting current and voltage

Science understanding

FOUNDATION | **STANDARD** | ADVANCED

1 Predict the voltages of each of the light globes in the circuits shown. Write the voltage next to each globe.

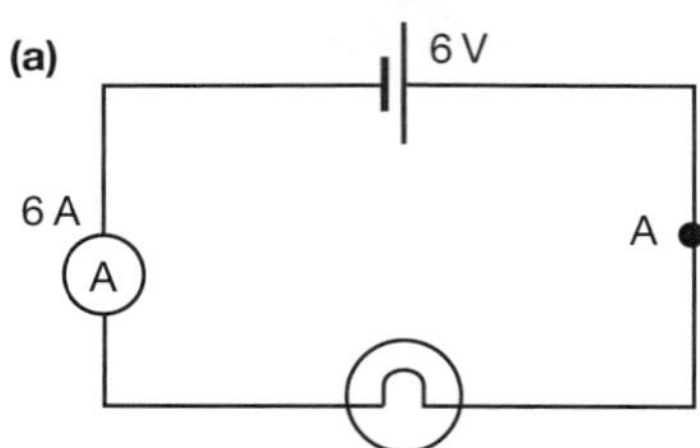

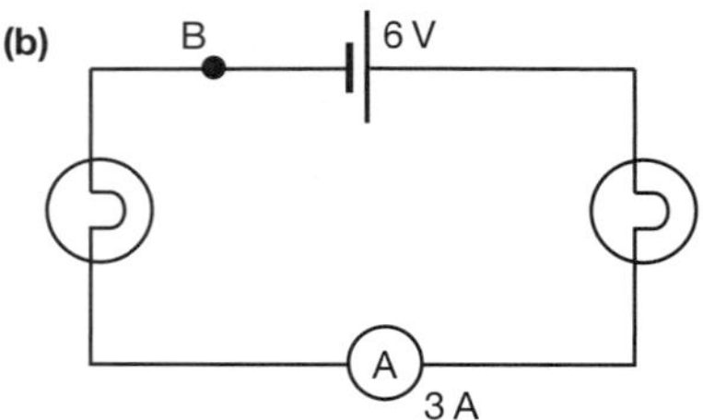

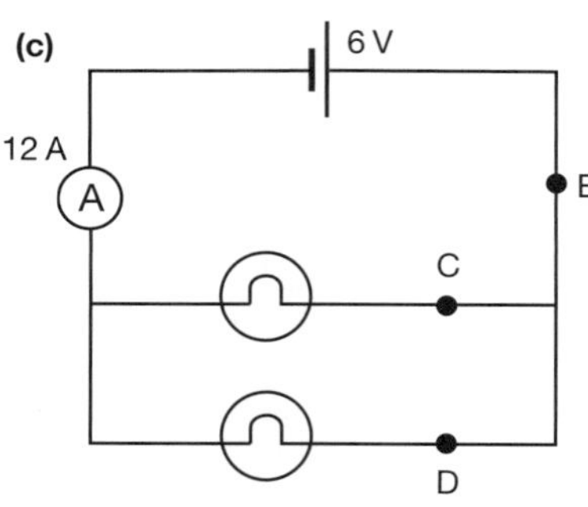

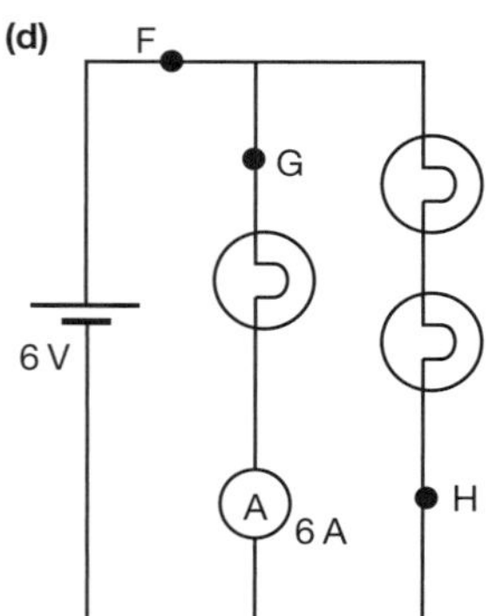

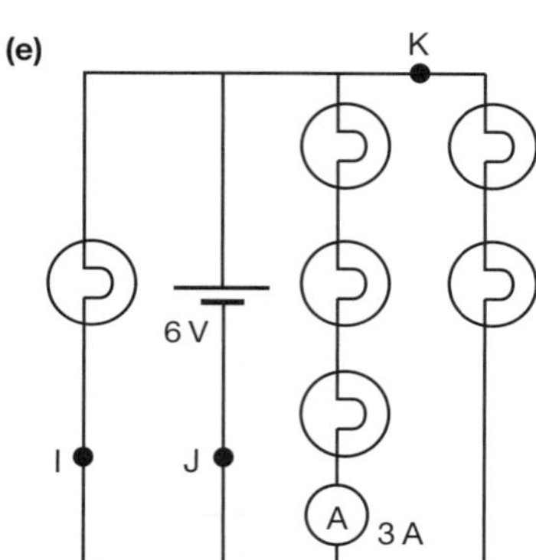

2 An ammeter is included in each circuit in question 1. From each ammeter reading, predict the current that is flowing through the points labelled A, B, C, D etc. in the above circuits.

Point A = __________ A

Point B = __________ A

Point C = __________ A

Point D = __________ A

Point E = __________ A

Point F = __________ A

Point G = __________ A

Point H = __________ A

Point I = __________ A

Point J = __________ A

Point K = __________ A

RATE MY UNDERSTANDING
Shade the face that shows your rating

6.7 Electrical safety

Science understanding

FOUNDATION	**STANDARD**	ADVANCED

Electricity is so common that we all take it for granted. However, it is also incredibly dangerous. For example, a circuit can quickly overheat and catch fire if you piggyback double-adaptors or powerboards. To avoid this, use separate plugs. Likewise, coiled or looped-up leads and extension cords can rapidly heat up, possibly melting their insulation and exposing live wires. Straighten them out first before using. Electricity is most dangerous if you accidentally become part of the circuit! If that happens, current may flow through you to the ground. Serious burns or electrocution can result.

Water can provide an easy route for electric current to flow through you to the ground. For this reason, never use an appliance or turn a switch on or off if you are wet. Dry yourself first. Likewise, never use electricity around swimming pools or filled basins or baths. For example, use your hairdryer in your bedroom, not the bathroom.

Never touch someone who has collapsed from electric shock, because they may still be part of the circuit! Turn off the power at the main switchboard and ring for an ambulance. Dial 000 from a landline phone. From a mobile, dial 000 or 112 or 106.

1. Conduct research into safe procedures when using electricity and electrical appliances. Prepare a table of 'Do' and 'Do not' procedures that can guide safe usage of electricity. Write and/or illustrate the safety procedures in the table below.

Safety procedures for using electricity and electrical appliances	
Do	**Do not**

2. List the main safety devices that give some protection from electric shock and electrocution.

6.7 Electrical safety

3 Explain why power leads should be straightened out and not left in loops.

4 Propose a reason why piggybacking double adaptors and powerboards off the one powerpoint can cause them to overheat.

5 Explain why putting a knife into a toaster while it is working is dangerous.

6 Explain why water is dangerous around electricity.

7 Imagine you have come across someone who has been electrocuted. You want to move them away from a live electrical cable but don't want to touch them because you might become part of the circuit too. Suggest ways in which you could safely get them away from the electricity.

8 What are the emergency phone numbers that you should use if you come across an emergency?

9 Assess your use of electricity at home. List ways in which you could improve your electrical safety.

RATE MY UNDERSTANDING
Shade the face that shows your rating

6.8 Comparing methods of power generation

Science inquiry skills

FOUNDATION | STANDARD | **ADVANCED**

Processing & Analysing | Evaluating | Communicating

Electricity can be generated in many different ways. Study Table 6.8.1. Most of the information from the table has been converted to a set of graphs shown in Figure 6.8.1.

Table 6.8.1

Considerations	Method of power generation				
	Coal	Nuclear	Wind	Hydro	Solar
A CO_2 emitted (tonne)	247 383	50 215	207	248	448
B Average life of plant (years)	30	35	25	40	25
C Cost of every megawatt hour of electricity produced (in 2007) (A$)	$40	$46	$42	$48	$245
D Cost to build (in 2007) (A$ millions)	$1590	$2145	$732	$236	$900
E Average energy output (MWh)	800	1 105	47	50	20
F Maximum energy output (MWh)	1 000	1 300	150	100	100

* 1 MW = 1 million watts

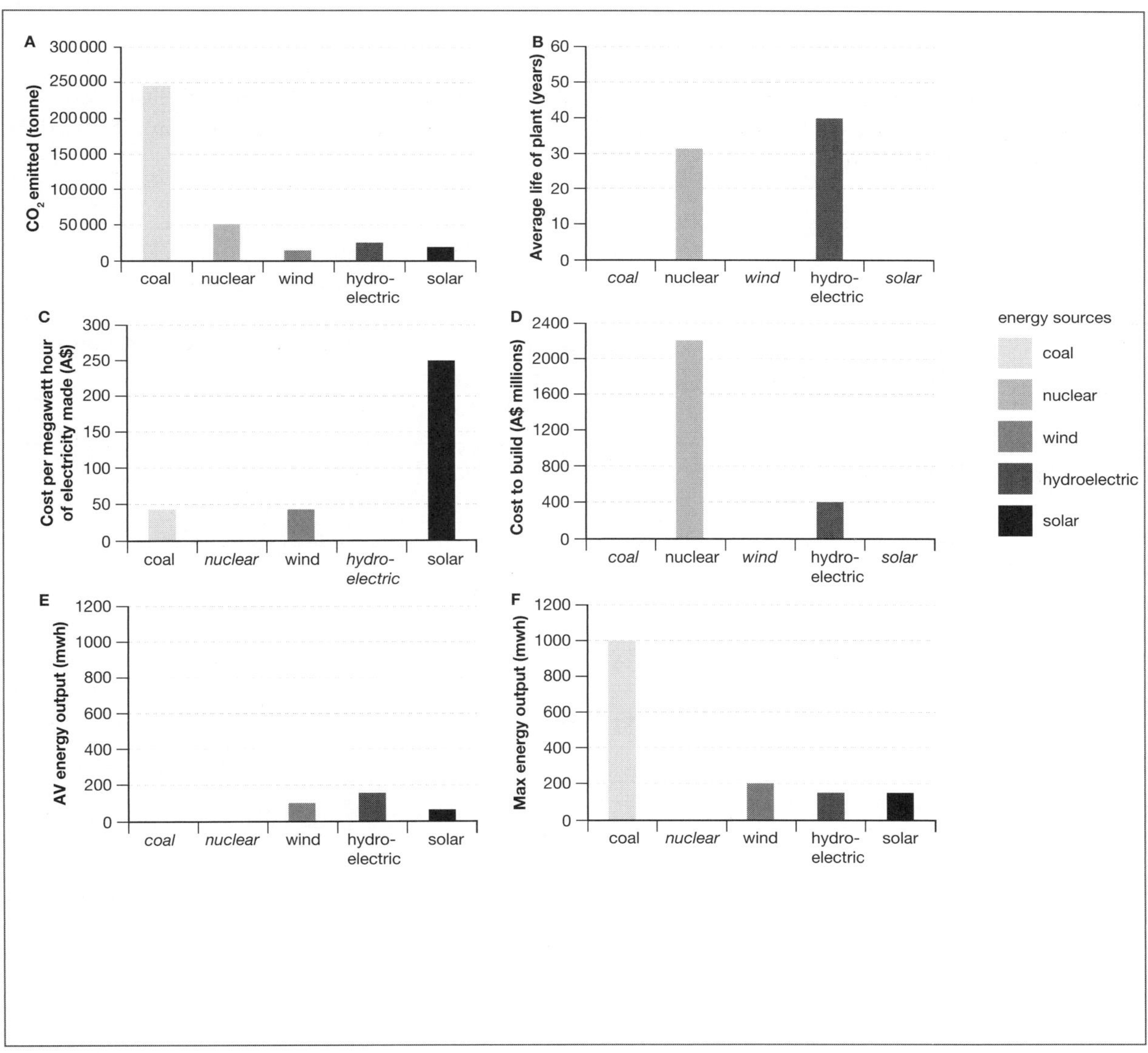

Figure 6.8.1

6.8 Comparing methods of power generation

1. Complete the Figure 6.8.1 graphs by referring to the table and constructing the 11 columns that are missing. Missing columns are indicated by italics on the horizontal axes.

2. List the power generation methods in order from:

   (a) cheapest per megawatt hour of energy produced to most expensive

   ____

   (b) least CO_2 emitted to the most CO_2 emitted

   ____

   (c) lowest average energy output to highest average energy output.

   ____

3. There is a large difference between the maximum energy output and average output from wind, hydro and solar plants. The difference is much smaller for coal and nuclear power plants. Propose reasons why.

   ____

   ____

   ____

4. Hydro, wind and solar power plants still contribute to carbon dioxide being released, even though none is released in the actual generation of electricity. Where do you think this carbon dioxide comes from?

   ____

   ____

5. **(a)** Use the information to assess which is the best method or range of methods of power generation for Australia.

   ____

   (b) Justify your answer.

   ____

   ____

   ____

   ____

   ____

RATE MY UNDERSTANDING
Shade the face that shows your rating

6.9 Literacy review

Science understanding

FOUNDATION	STANDARD	ADVANCED

1 Use words from the box to complete the sentences below. Use each word only once.

alternating	ammeter	ampere	biomass	conducts	direct	earth
electrocution	electrolyte	electromagnetism	electrons	fuse	geothermal	neutral
ohm	series	solar	solenoid	transformer	turbine	voltmeter

(a) Electricity is a type of energy that involves the movement of negatively charged particles called ______________ .

(b) Electricity flows through a circuit which is produced by joining a number of components together. When components are arranged one after another in a line, this is called a ______________ circuit.

(c) The unit used for measuring electrical current is ______________ .

(d) There are two main types of current. A current that shuffles back and forth repeatedly is called an ______________ current. A current that travels in one direction only is called a ______________ current.

(e) A ______________ is a type of large generator used to help create electricity.

(f) Electricity can be produced by using renewable resources such as ______________ heat, which is heat from the Earth, ______________, which uses energy from the sun and ______________, which involves using biological material.

(g) A ______________ is used to increase or decrease the voltage in a circuit.

(h) The relationship between electricity and magnetism is called ______________ .

(i) A common type of electromagnet is a ______________, which is a long coiled piece of wire.

(j) A solution or liquid that ______________ electricity is known as an ______________ .

(k) The instrument used to measure current is an ______________ and an instrument to measure voltage is a ______________ .

(l) Resistors are used to measure resistance in a circuit. The unit for resistance is __________ .

(m) The name given to the green and yellow wire in a power lead is ______________ , and the name given to the blue wire in a power lead is ______________ .

(n) A ______________ is a thin strip of metal that melts when too much current runs through a circuit, which causes a break in the circuit. Misuse of electricity could result in ______________ .

RATE MY UNDERSTANDING
Shade the face that shows your rating

6.10 Thinking about my learning

Tick the box that best reflects your understanding of each concept for this chapter on electricity.

The main ideas for this topic	I still need help with this	I understand this	I understand this and can explain this concept to other students
I am familiar with the structure of an atom and the charge on each particle.			
I understand the difference between static electricity and current electricity.			
I know the symbols for the components of a simple circuit.			
I am able to accurately draw an electric circuit using the correct symbols.			
I am confident about constructing a simple circuit.			
I understand the difference between current and voltage.			
I am able to take an accurate reading of different meters used to measure volts and amperes.			
I understand how resistance affects the current in a circuit.			
I understand the basic operation of a photovoltaic (solar) cell.			
I know the difference between conductors and insulators.			
I can recognise the difference between a series and parallel circuits.			
I understand the need to have safety measures built into electrical circuits.			
I understand how energy is converted from one form to another when electricity is generated from different sources.			
I have some understanding of how magnets and electricity work together.			
I am able to interpret data and apply my knowledge to unfamiliar information.			

CHAPTER 7

Body coordination

7.1 Knowledge preview

Science understanding

FOUNDATION | **STANDARD** | ADVANCED

1 Answer TRUE or FALSE for each of the following statements.

Statements	True or False
Hormones are special chemicals that circulate in the bloodstream.	
Nervous tissue is made up of special cells called axons.	
Nervous tissue is only found in specialised parts of the human body.	
Oestrogen is a hormone involved in controlling the reproductive cycles of mammals.	
The sensory organs contain specialised cells that are part of the nervous system.	
Only the nervous system controls the functioning of our body.	
Nerve cells send messages by electrical signals.	
Hormones act much more quickly than nerve signals.	
Both hormones and nerve cells respond to a specific stimulus.	
Adrenaline is a hormone that allows us to respond more quickly to danger.	

2 Below is a list of some terms involved in the control of our body systems. Circle the ones you have heard about or are familiar with. Do not be concerned if some are unfamiliar, because you will learn about them in this chapter.

adrenaline	cortisol	diffusion	endocrine system	endothermic
enzymes	homeostasis	hormones	insulin	metabolism
neuron	nutrients	oestrogen	progesterone	testosterone

3 Table 7.1.1 is a list of stimuli our body responds to, the organs that respond to the stimulus and the response we have to that stimulus. The columns are not in the correct order. Look at the lists and re-write them correctly in the flow charts below.

Table 7.1.1

Organs, stimuli they respond to and the type of response		
Stimulus	**Organ**	**Response**
increased CO_2	heart	lowered blood sugar
big meal	kidney	increased breathing rate
sudden noise	lungs	concentrated urine
bright light	eye	rapid heart rate
dehydration	pancreas (insulin)	pupil contracts

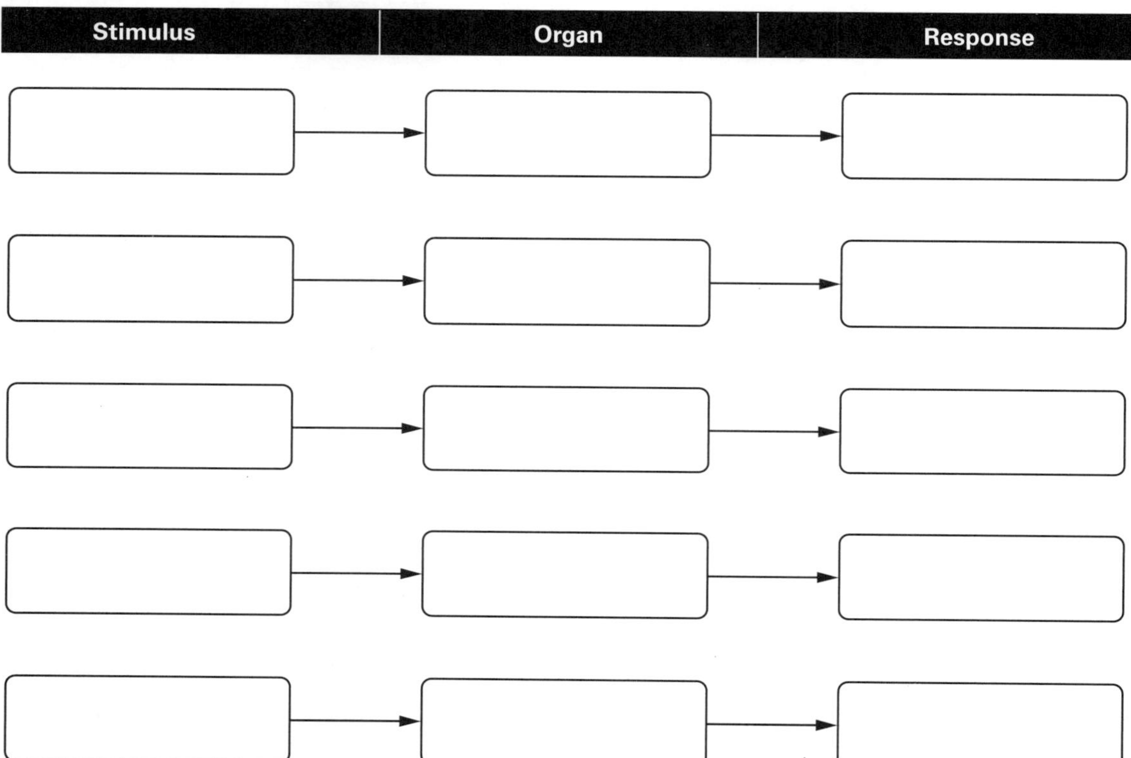

7.2 Kidney function

Science understanding

FOUNDATION | STANDARD | **ADVANCED**

Filtering wastes

The human body has two kidneys. The kidneys filter about 230 litres of fluid every 24 hours, of which about 225 litres is returned to the blood stream and 5 litres is urine we excrete. The urine is collected in the bladder and can be held there from 1 to 8 hours. Each kidney has a million nephrons which are microscopic filters. The function of the nephrons is to filter your blood to remove harmful wastes. The nephrons also ensure that useful substances are returned to the blood and not excreted with the wastes. Figure 7.2.1 shows the structure and function of a nephron.

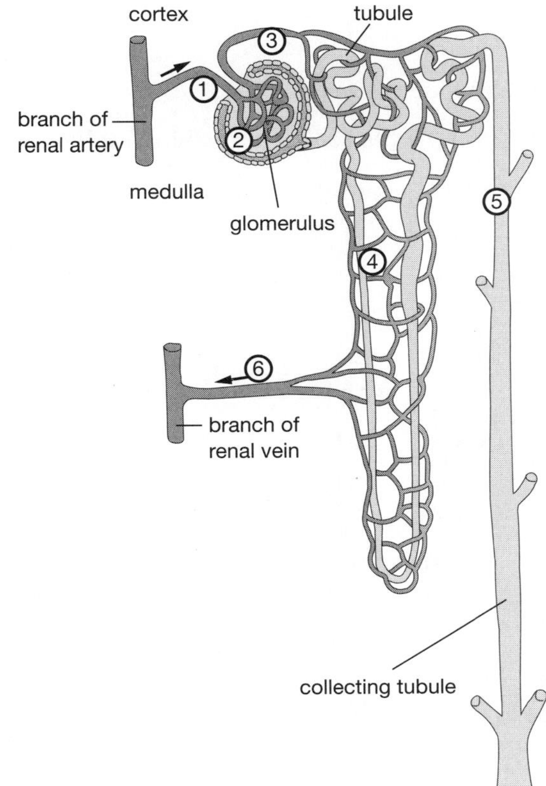

1 Blood enters the glomerulus in the kidney from the renal artery.

2 High blood pressure in the glomerulus forces glucose, water, urea and salts through the capillary walls and into the tubule.

3 Protein molecules and blood cells are too large to pass through the capillary wall so they remain in the blood. Blood leaves the glomerulus and flows to the renal vein.

4 The glucose, water and salts pass through the tubule. They are absorbed back into the blood by nearby capillaries so blood becomes the right concentration.

5 Urea and other unwanted wastes stay in the tubule and go into the bladder to be excreted from the body.

6 Blood with lower levels of salt and urea flows out of the kidney into the renal vein.

Figure 7.2.1 The structure and function of a nephron

1 What are nephrons?

____

____

(2) Explain how the endocrine system is involved in excretion.

____

____

(3) Explain how the circulatory system is involved in excretion.

____

____

7.2 Kidney function

(4) Predict how excretion would be affected if the:

(a) kidneys were further from the heart

____

(b) tubule was shorter

____

(c) tubule was not surrounded by so many capillaries.

____

5 On this simplified diagram of a nephron (Figure 7.2.2):

(a) use green arrows to show the direction filtered blood moves

(b) use blue arrows to identify the pathway the materials filtered from the blood follow

(c) use red arrows to identify where useful materials are returned to the blood, and the pathway they then follow.

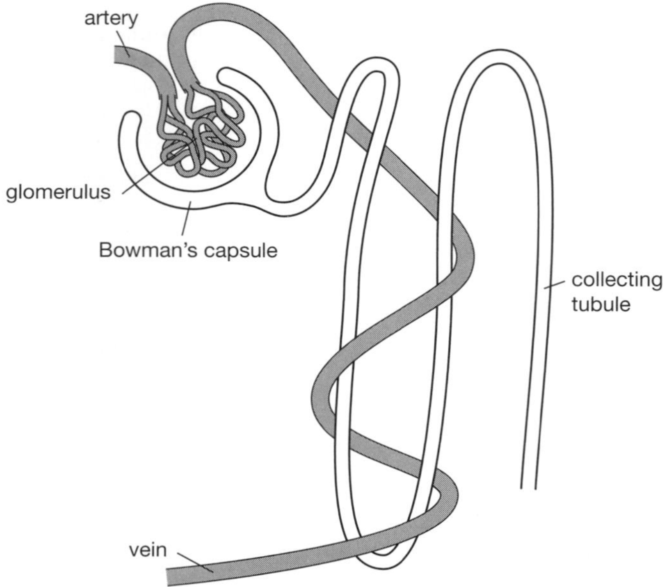

Figure 7.2.2 Simplified diagram of a nephron

6 List four materials returned to the blood from the filtered material.

____

____

RATE MY UNDERSTANDING
Shade the face that shows your rating

7.3 The nervous system

Science understanding

FOUNDATION | STANDARD | ADVANCED

1 Select the correct terms from the box to label the parts of the nervous system (Figure 7.3.1) and the structure of nerves (Figure 7.3.2).

Parts of the nervous system

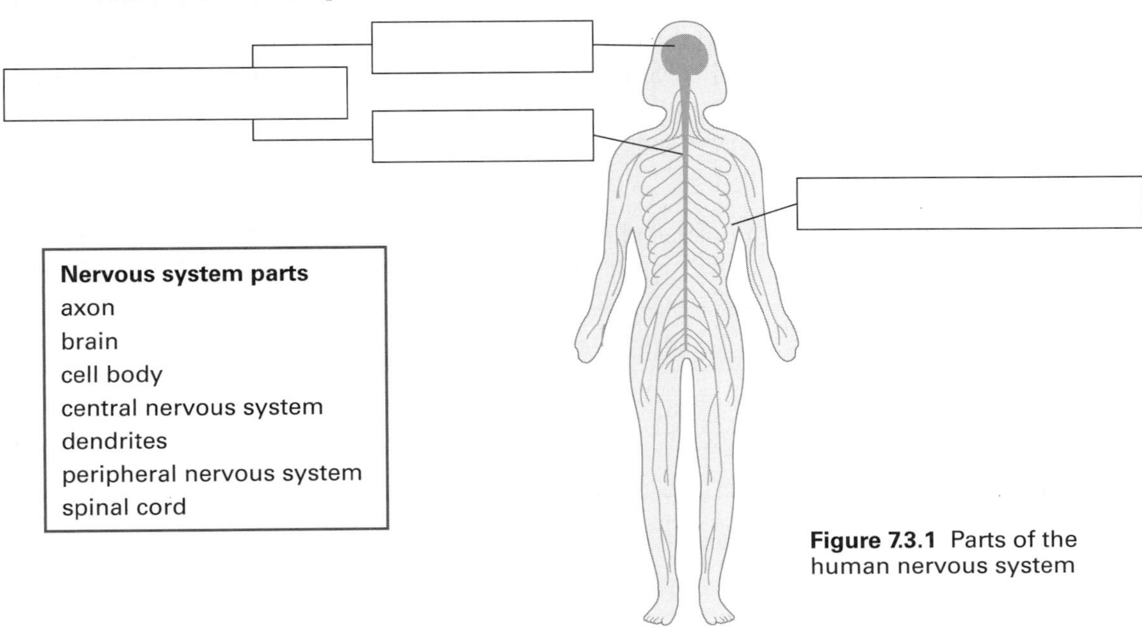

Figure 7.3.1 Parts of the human nervous system

The structure of nerves

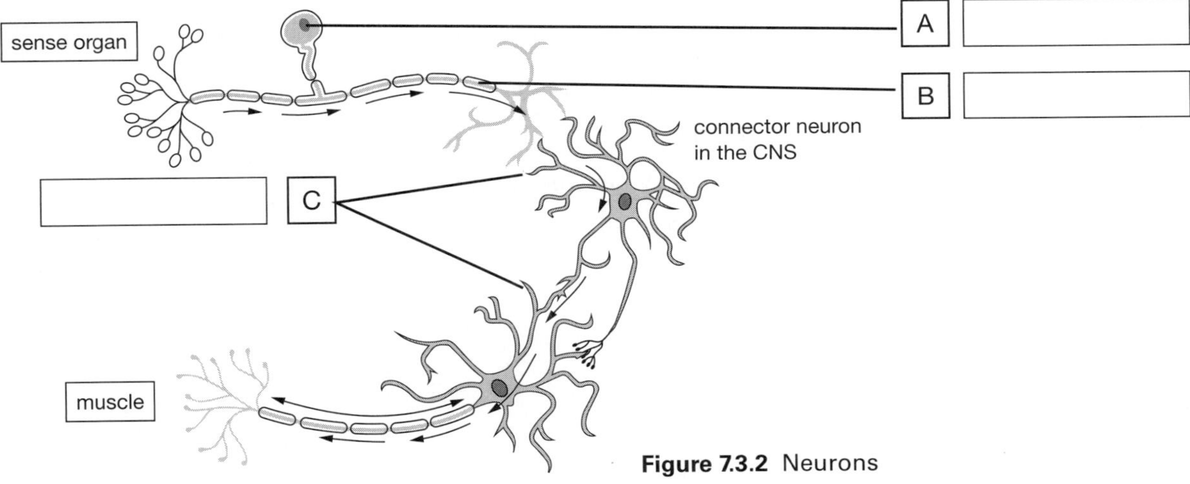

Figure 7.3.2 Neurons

2 Look at Figure 7.3.2, The structure of nerves, and complete the following labels.

(a) Use arrows to show the direction of the nerve impulse.

(b) Label the sensory neuron and the motor neuron.

(c) Identify and circle two positions where there are synapses.

RATE MY UNDERSTANDING
Shade the face that shows your rating   

7.4 Reflexes

Science understanding

FOUNDATION | **STANDARD** | ADVANCED

Reflexes are messages that travel from where they are received, along a sensory neuron to the spinal cord. Within the spinal cord, a relay neuron transmits the message straight back along motor neurons to the first place they were received. This causes a response called a reflex action.

Because reflexes do not have to pass all the way up to the brain and back, they are very fast. Reflexes help protect you from danger. Examples of reflexes are: pulling your hand away from a hot object, blinking your eyes to stop something getting into them, and pulling your foot away after standing on a sharp object. A message is sent to the brain shortly afterwards. Only then can the brain register pain.

1 Compare a reflex action with normal stimulus–response reactions of the nervous system.

____

____

____

____

(2) Explain why this makes the reflex action so much faster.

____

____

____

(3) Use Figure 7.4.1 to demonstrate the path of a reflex action.

(a) Identify the pathway of motor neurons in red.

(b) Identify the relay neuron in green.

(c) Identify the pathway of the sensory neurons in blue.

(d) Use arrows to demonstrate the direction of the stimulus.

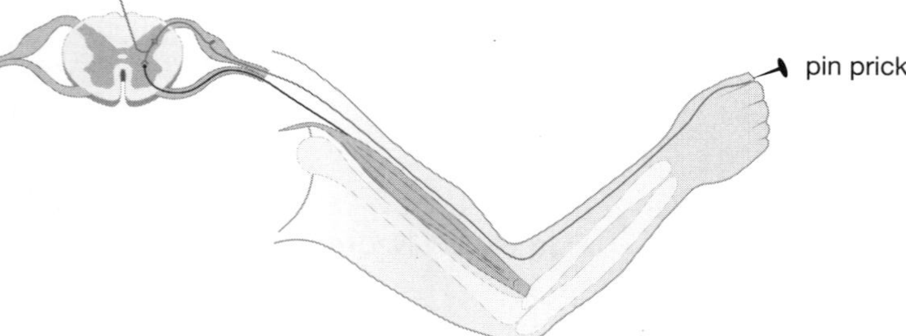

Figure 7.4.1

4 Explain why you still feel the pain even after you have moved away from the cause of the pain (stimulus).

____

____

____

RATE MY UNDERSTANDING
Shade the face that shows your rating   

7.5 Sweating

Science understanding

FOUNDATION	STANDARD	ADVANCED

dehydration (*n*) a lack of water or fluid
duct (*n*) tube or pipe
excess (*adj*) too much
regulate (*v*) to control

Your skin and sweat glands are part of the excretory system. Sweat glands are located just under the skin and are found all over your body. Each person has 2 million to 4 million sweat glands. Sweat is essential for survival. Sweating is like a coolant that stops the body from overheating. Excess sweating means your body has lost large amounts of water and salts, resulting in dehydration. The body needs salts to regulate the water content in the blood. It is therefore important to drink water after exercise or when temperatures are very high.

The sweat gland is composed of two parts (Figure 7.5.1). The coiled part is where sweat is produced. The duct leading to a pore on the skin surface is the tube through which sweat is excreted from the body. The fluid that produces sweat comes from the spaces between body cells. This fluid, which is mainly water, has high concentrations of sodium and chloride and low concentrations of potassium and urea. As fluid travels up the sweat duct it changes, depending on conditions such as temperature. The diagram demonstrates these changes when conditions are hot and when they are cold.

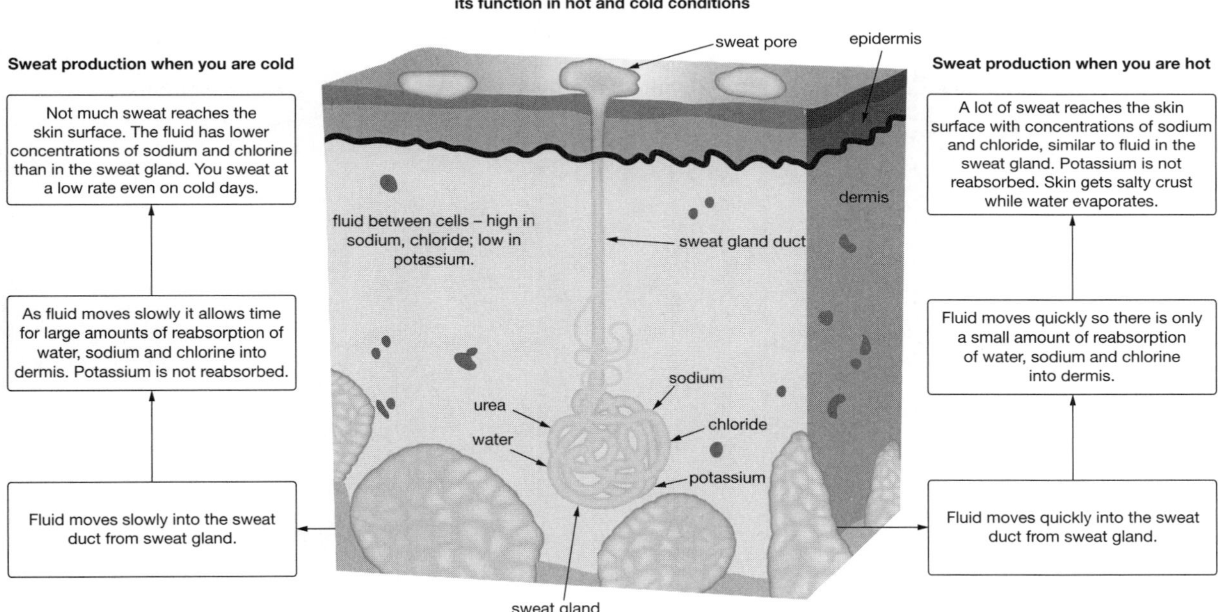

Figure 7.5.1 The human sweat gland

1 List the excretory products that leave the body through the skin.

____

2 Explain where the fluid that forms sweat comes from.

____

____

7.5 Sweating

(3) **(a)** Do you think that your skin would taste saltier on hot days or cool days?

____

(b) Justify your answer.

____

____

4 Explain why sweating is necessary.

____

____

5 Describe a situation where excessive sweating could be harmful.

____

____

6 Compare sweating and the functioning of sweat glands when a person is hot and when they are cold. Show the similarities, differences and common features on the Venn diagram below.

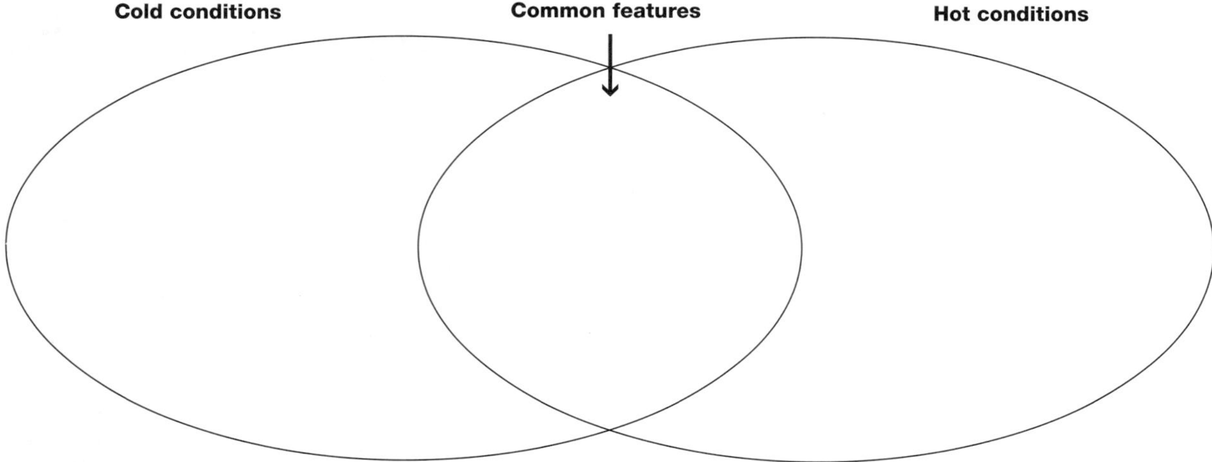

RATE MY UNDERSTANDING
Shade the face that shows your rating   

7.6 Reaction time

Science understanding

FOUNDATION | **STANDARD** | ADVANCED

Reflex actions happen very fast, compared with other actions by the body in response to danger.

You are travelling very quickly on your skateboard or bike and suddenly see someone walk in front of you. You have to make judgements about how far away the person is, what actions you can take to avoid them, which action is the best in the situation, and whether you have the time and skills to carry it out. You then have to follow through on your decision.

The time it takes for you to do all these things is your reaction time.

Experiments show that when you are attentive you have a reaction time of between 0.2 and 1.5 seconds. However, police estimates show that the average driver takes up to one second to react in an emergency.

Once you have reacted, you have to bring your skateboard or bike to a stop. The distance you travel before stopping is the reaction distance. This distance depends on the speed at which you are travelling and your reaction time.

To calculate reaction distance:

reaction distance = speed × reaction time

$$d = v \times t$$

Where: d is the reaction distance measured in metres

t is the reaction time measured in seconds

v is the speed of travel measured in metres per second

1. Calculate the reaction distance for each situation in Table 7.6.1. Use the space below the table for your calculations.

Table 7.6.1

Reaction time and speed						
v Speed (m/s)	1.7	4.2	5.6	11.1	13.9	27.8
Equivalent Speed (km/h)	6	15	20	40	50	100
What travels at this speed	person walking	person running	recreational cyclist	car in school zone	car in residential street	car on highway
t Reaction time (s)	1	0.7	0.6	0.5	0.5	0.7
d Reaction distance (m)						

7.6 Reaction time

(2) Consider what would happen if a driver was distracted and took longer to react. Calculate the reaction distances with the slower reaction times in Table 7.6.2. Use the space below the table for your calculations.

Table 7.6.2

Reaction time and speed						
v Speed (m/s)	11.1	11.1	13.9	13.9	27.8	27.8
Equivalent Speed (km/h)	40	40	50	50	100	100
What travels at this speed	car in school zone	car in school zone	car in residential street	car in residential street	car on highway	car on highway
t Reaction time (s)	1.5	2	1.5	2	1.5	2
d Reaction distance (m)						

(3) Why are speed limits lower (40 to 50 km/h) in school zones and in residential streets?

4 What factors do you think would reduce the attentiveness of drivers in the following situations?

(a) in a school zone

(b) in a residential street

(c) on a highway.

RATE MY UNDERSTANDING
Shade the face that shows your rating

7.7 Diseases of the nervous system

Science understanding

FOUNDATION | **STANDARD** | ADVANCED

Diseases of the nervous system affect the ability of the body to control movement and other functions. Parkinson's disease and cerebral palsy are two diseases of the nervous system.

Parkinson's disease

The muscles of a person with Parkinson's disease are often tight, causing their voluntary movements to be slower than normal. The person may also experience uncontrollable tremors. These changes are caused by a reduction in the number of neurons that make a chemical called dopamine. Dopamine is a neurotransmitter, a chemical that carries messages from one cell to another. Dopamine enables human muscle movement to be smooth and coordinated. When approximately 60% to 80% of the dopamine-producing neurons are damaged, too little dopamine is produced and the symptoms of Parkinson's disease appear.

Scans of the brains of people with Parkinson's disease reveal the loss of dopamine-producing neurons in part of the midbrain. They also show accumulations (build up) of a certain protein. This protein is produced as the neurons degenerate. It is, in effect, waste that the body cannot get rid of as quickly as it is produced.

The degeneration of the neurons and the evidence of accumulated proteins are both used in the diagnosis of Parkinson's disease. However, they can only be detected once the disease has progressed far enough to produce symptoms. The person may have had the disease for more than 20 years by that stage.

A current theory is that the early stages of Parkinson's disease affect the olfactory bulb, the part of the brain that controls your sense of smell. There is an increasing amount of evidence that loss of a sense of smell and sleep disorders are present in people with Parkinson's disease long before their muscles are affected. Research is focusing on these early symptoms in the search for ways of slowing or stopping the progress of the disease.

1 Describe the role of dopamine in the body.

____

2 A reduction in the amount of dopamine results in the slowness of movement that is a symptom of Parkinson's disease. Propose a reason why.

____

____

3 Summarise the evidence that is traditionally used in the diagnosis of Parkinson's disease.

____

____

(4) **(a)** Summarise the possible link between a loss of the sense of smell and Parkinson's disease.

____

____

(b) Explain the benefit of making this link to people with Parkinson's disease.

____

____

7.7 Diseases of the nervous system

Cerebral palsy

Cerebral palsy is the most common form of childhood physical disability, affecting about 34 000 Australians. It is not contagious, nor is it a disease passed from parents to children. It is a disability caused by changes or injury to the brain as it is developing. The injury may occur before birth. It can also be caused by a lack of oxygen reaching the brain during birth or soon after birth.

The word cerebral refers to the brain and palsy means weakness or paralysis. Cerebral palsy is a disability that affects movement of the body. The muscles in your body have tension and it is this tension that allows you to stand and to move your limbs. The tension in your muscles is a response to messages from your brain. In a person with cerebral palsy, the messages sent from the brain to the muscles may be sent at the wrong time or may be sent to the wrong place. The muscles that receive the messages become confused. The result may be extra tension in the muscle causing shaking and jerking of the limbs. Alternatively, muscle tension may be reduced causing the muscles to be flaccid (low tension).

Cerebral palsy affects different people in different ways. Some people:

- have minor problems with fine motor skills and just appear to be clumsy
- are affected on only one side of their body
- have only their legs affected or less commonly have only their arms affected
- are totally physically dependent because both arms and legs along with the muscles of the face and mouth are affected.

(5) Explain why the name cerebral palsy is also a description of the disease.

____

____

(6) Cerebral palsy can be described as a group of disorders rather than a single disability. Propose a reason why this description is reasonable.

____

____

7 Compare Parkinson's disease and cerebral palsy.

Similarities	Differences

RATE MY UNDERSTANDING
Shade the face that shows your rating

7.8 Bionic ear and eye

Science as a human endeavour

FOUNDATION	STANDARD	ADVANCED

1. Use figures 7.8.1 and 7.8.2 to compare the human ear and a bionic ear; then complete the table below. Identify which parts of the human ear are equivalent (similar/the same) to the parts of the bionic ear. Where you do not think there is a direct equivalent, and why?

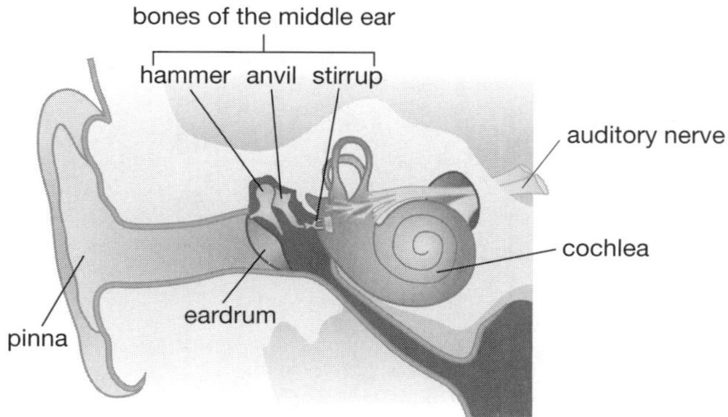

Figure 7.8.1 Structure of the human ear

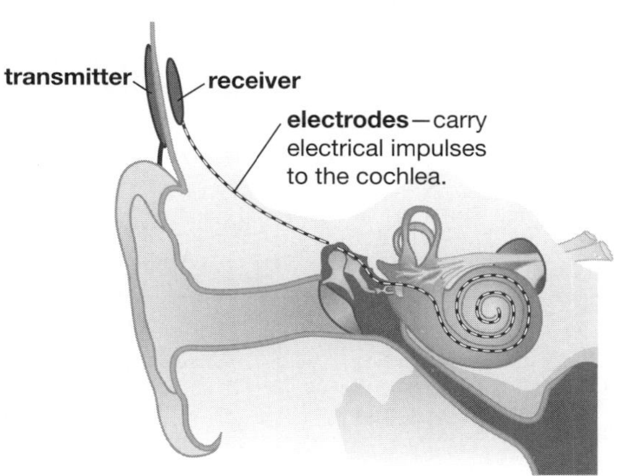

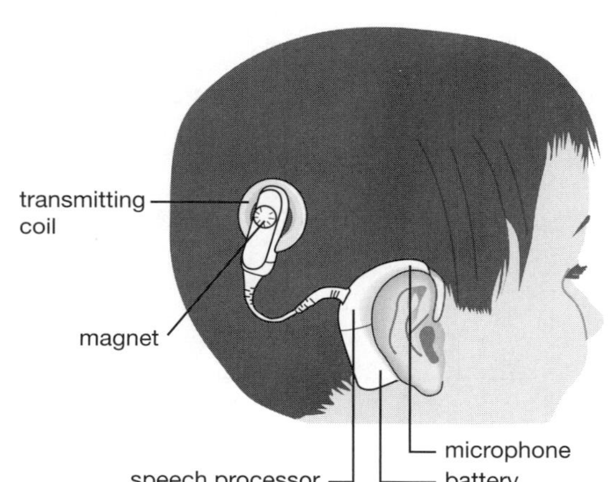

Figure 7.8.2 Structure of the bionic ear

Part of bionic ear	Equivalent in human ear	Justification
microphone		
speech processor		
transmitting coil		
receiver		
electrode		

7.8 Bionic ear and eye

(2) Although the bionic ear (Figure 7.8.2) and bionic eye (Figure 7.8.3) have different functions, they have some things in common.

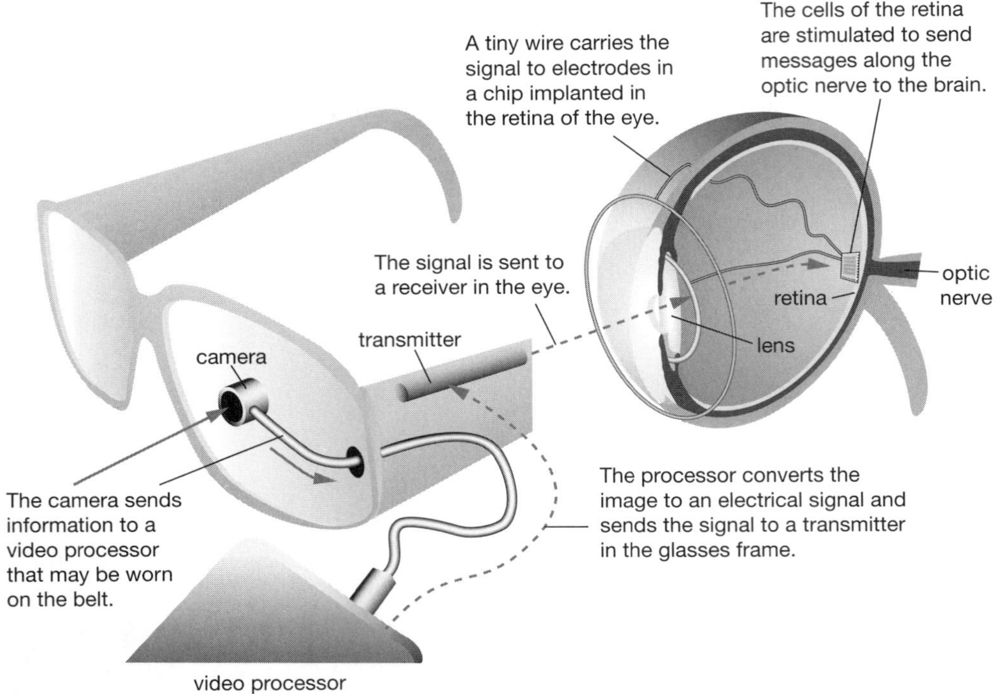

Figure 7.8.3 How a bionic eye works

Compare the bionic ear and the bionic eye.

____

____

____

3 What do you think are some of the short-term and long-term advantages of a young child being fitted with a bionic ear? Complete the following table.

Short- and long-term advantages of a bionic ear	
Short-term advantages	**Long-term advantages**

4 What do you think are some of the short-term and long-term advantages for the patient in having a bionic eye? Complete the following table.

Short- and long-term advantages of a bionic eye	
Short-term advantages	**Long-term advantages**

RATE MY UNDERSTANDING
Shade the face that shows your rating

7.9 Temperature control

Science understanding

FOUNDATION | STANDARD | **ADVANCED**

Feedback systems

The control of temperature is a feedback system. There are three main parts to a feedback system: sensor, controller and effector.

> **activate** (*v*) to make something work, to turn on
> **detect** (*v*) to find out, to discover

Automatic heating and cooling of a building depends on a feedback system. Once the thermostat has been set at a particular temperature, the thermometer (the sensor) detects (senses) the temperature. If the temperature is below the set temperature, the switch (controller) is activated and the heater (effector) starts working to raise the temperature. The rise in temperature feeds back to the thermometer (the sensor). When the set temperature is reached, the switch turns the heater off.

This information can be summarised in a simple diagram (Figure 7.9.1).

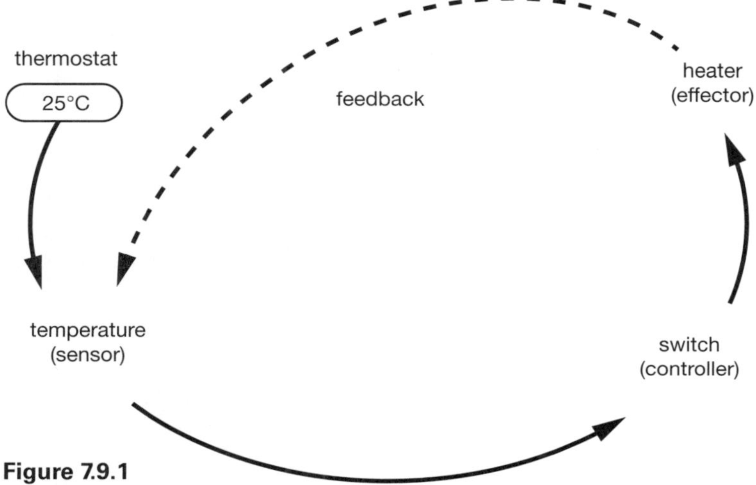

Figure 7.9.1

1. Imagine the thermostat is still set at 25°C. Construct the feedback system that would cool the room if the sensor detected that the room was too warm (35°C).

> **HINT**
> Think about what the effector would be instead of the heater.

7.9 Temperature control

Temperature control in the body

The temperature control system of the body is more complex.

2. Describe what would happen to the body's feedback system when the core temperature (the temperature within the body) starts to rise.

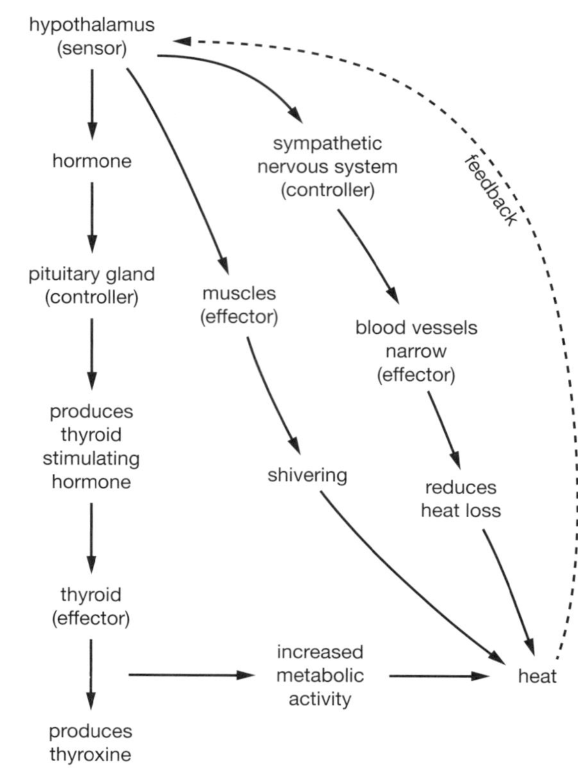

Figure 7.9.2 The body's feedback system when the core temperature falls

Your skin also has temperature sensors. These temperature sensors send nerve impulses to the brain. The brain 'tells' the body to react in the same ways as a stimulus from the hypothalamus. The brain also causes the hypothalamus to become more sensitive to changes in core temperature.

3. Discuss the advantages of having two types of temperature sensors—the receptors in the skin and the hypothalamus. Next, select the correct phrases from the list and write them in the gaps in the sentences below.

the temperature of your blood is changed there is a change in the external temperature alert your brain to the changes help you maintain a constant temperature behave in ways to help maintain your body temperature

The receptors in the skin react immediately when ______________________.

They ______________________. You can then ______________________. The hypothalamus detects changes later when ______________________ rather than just your surface temperature. It will cause involuntary reactions to ______________________ when you may already be used to the temperature of the environment.

RATE MY UNDERSTANDING
Shade the face that shows your rating

7.10 Literacy review

Science understanding

FOUNDATION	STANDARD	ADVANCED

Recall your knowledge of human body systems by matching the word in the left-hand column with its correct meaning in the right-hand column.

Word	Meaning
autonomic nervous system	the brain and spinal cord
brain stem	a chemical message released by the axon when the signal reaches the end of a neuron
catalyst	a muscle or gland that puts the message into effect
central nervous system	branch from the cell body that receives messages from other neurons
dendrite	a portion of the brain that constantly checks the internal environment of the body
diffusion	special cells that detect stimuli
effector	system controlling involuntary actions such as heartbeat
endocrine glands	the nerves that carry messages to and from the central nervous system and other parts of the body
hormones	movement of particles of a substance from an area of high concentration to an area of low concentration
hypothalamus	glands that produce hormones
receptors	chemical substances that act as messengers in the body
mitochondria	substance that speeds up the rate of reaction without being used up in the process
neurotransmitter	part of the brain that controls vital functions such as breathing
peripheral nervous system	the space between neurons
ribosomes	structures where proteins are manufactured
synapse	organelles that are the site of cellular respiration

RATE MY UNDERSTANDING
Shade the face that shows your rating

7.11 Thinking about my learning

1 Tick the box that best matches your understanding of this chapter.

Knowledge statements	I need help with this area of knowledge	I am confident with my knowledge and understanding
I know the difference between the central nervous system and the peripheral nervous system.		
I can label and name the parts of a neurone.		
I know the different parts of the central nervous system and the peripheral nervous systems.		
I understand what the somatic and autonomic systems do.		
I understand how a simple reflex arc works and how a signal is transmitted.		
I know the major organs of the endocrine system.		
I understand how the endocrine system sends a signal.		
I can name some hormones of the endocrine system.		
I understand the concept of homeostasis.		

2 The completed flow chart below shows the major stages in maintaining your body temperature but the steps are in the wrong order.

(a) Jumbled order of maintaining body temperature steps

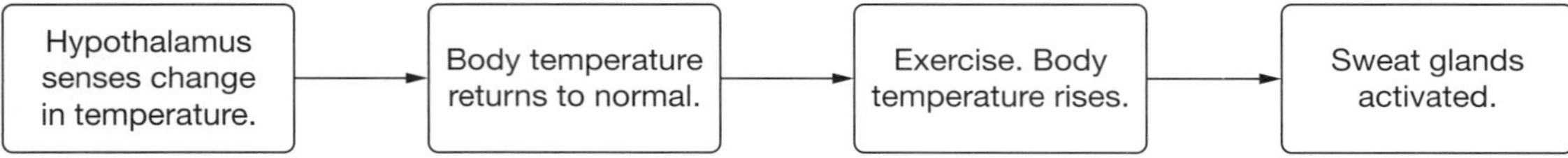

(b) Re-write the steps in maintaining body temperature in the flow chart below, in the correct order.

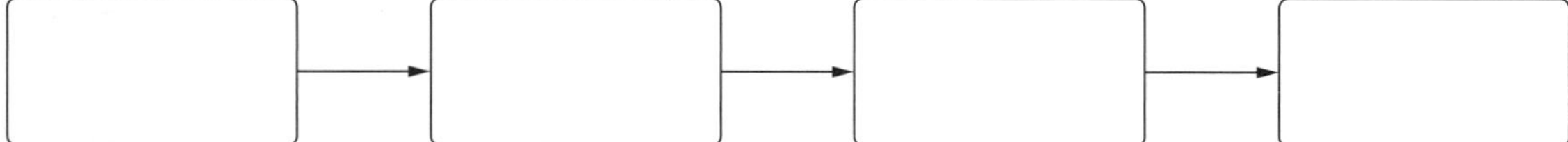

3 Complete your own flow chart to show how the body responds when blood sugar levels are low.

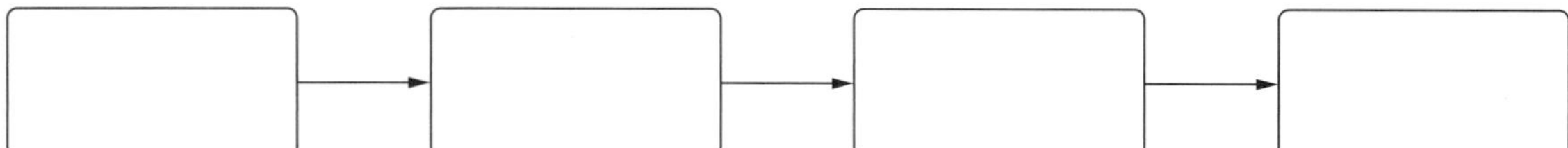

CHAPTER 8

Disease

8.1 Knowledge preview

Science understanding

FOUNDATION	STANDARD	ADVANCED

1 Diseases are conditions that can make you sick and stop your body from functioning normally. Diseases can be classified as infectious or non-infectious. Complete the following sentences by either writing 'infectious' or 'non-infectious'.

(a) Diseases that cannot be passed from one person to another are called ____________________

(b) Diseases that can be passed from one person to another are called ____________________

2 Classify the diseases in the box below as infectious or non-infectious by writing them in the correct column in the table.

AIDS	Alzheimer's disease	appendicitis	arthritis	athlete's foot (tinea)
breast cancer	chickenpox	cholera	diabetes	heart disease
influenza	malaria	measles	melanoma	multiple sclerosis (MS)
osteoporosis	poliomyelitis	rabies	syphilis	whooping cough

Infectious disease	Non-infectious disease

3 In Australia we are able to control many diseases that are major problems in some other countries. We have good hygiene measures, access to medicines and ways of boosting our immunity. Write the words listed in the box into the table so that they correspond to the correct health measure.

antibiotics	anti-malarial drugs	antiseptic creams	anti-viral drugs	cold-sore cream
disinfectants	exercise	hand washing	insecticides	pasteurisation
quarantine	refrigeration	using tissues	vaccination	washing food

Hygiene measures	
Medicines	
Immunity	

8.2 Growing bacteria

Science inquiry skills

FOUNDATION	STANDARD	ADVANCED

Communicating

Bacteria reproduce by one cell dividing into two. In ideal conditions, one bacterium can divide every 20 minutes. Imagine you left milk out of the fridge when you went to school, and there was one bacterium in the container.

1. Use your calculator to calculate the number of bacteria you would have at the end of each hour for the next 8 hours. Complete the second row of the table below.

Growth of bacteria with time									
Time (h)	0	1	2	3	4	5	6	7	8
Number of bacteria	1								
Length of vertical axis (cm)	0.001								

2. Construct a graph showing the increase in the number of bacteria for the first 4 hours.

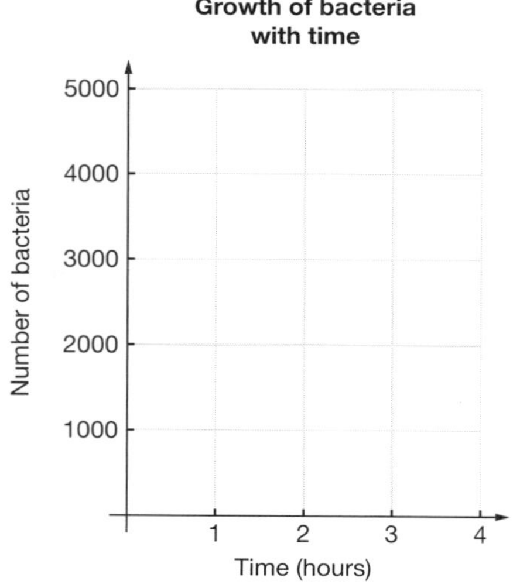

3. If 1 cm = 1000 bacteria, calculate the length of the vertical axis needed to plot the numbers for the remaining 4 hours. Add your calculations to the third row of the table above.

4. Describe the shape of the graph.

____

5. If food was infected with bacteria that cause food poisoning, it is unlikely that there would be only one bacterium to begin with. There would be thousands.

   Calculate the number of bacteria on food after 8 hours if 1000 bacteria were present to begin with. Assume ideal conditions.

____

____

RATE MY UNDERSTANDING
Shade the face that shows your rating

8.3 Traditional and modern medicine

Science as a human endeavour

FOUNDATION	STANDARD	ADVANCED

Read the following article and then answer the questions.

Indigenous medicine meets biomaterials development

Australian native plants, traditional Aboriginal culture, biotechnology and improved healthcare all come together in a uniquely Australian good news story.

Professor Hans Griesser, Doctor Susan Semple and collaborators at the University of South Australia have identified a group of new antibacterial chemicals called 'serrulatane diterpenes' from the resin and leaves of many species of *Eremophila,* and are using them to prevent bacterial growth on medical implants. Over 216 species of *Eremophila* have been found across arid regions of inland Australia. Aboriginal people have used extracts from several of them to treat skin sores and sore throats for generations. It was this pattern of use that had suggested an antibacterial action, and this was subsequently confirmed in the laboratory. All the active compounds isolated were able to prevent the growth of dangerous multi-drug resistant strains of bacteria, such as methicillin-resistant *staphylococcus aureus* (MRSA), which are often the cause of serious hospital infections.

By employing specially designed adhesive layers, the active compounds isolated from the plants were joined to the surfaces of polymer and ceramic implant materials ... retaining their antibacterial activity in the process.

[Specific tests] were used to confirm the presence of ... specific diterpenes on the surfaces, and ... the linkage chemistry (joining chemistry) was as it should be. Culturing bacteria on treated and untreated surfaces was the next step and ... bacteria were very unhappy on the diterpene-coated surfaces. [*continued on next page*]

active (*adj*) responsible for the action of something

antibacterial (*adj*) something that destroys bacteria

arid (*adj*) dry

biomaterial (*n*) a material that replaces or adds to a part of the biological process

collaborator (*n*) a person you work with, a partner

medical implant (*n*) a device inserted into the body

serrulatane diterpenes (*n*) naturally occurring organic compounds

subsequently (*adv*) afterwards

1 Serrulatane diterpenes are complex chemicals. Describe the characteristics they have that are of great interest to scientists.

____

2 What diseases did Indigenous Australians treat with *Eremophila* extracts?

____

3 Explain the term active compound.

____

4 Complete the sentences to outline the process used to confirm the antibacterial properties of the active compounds in *Eremophila*.

(a) The active compounds of *Eremophilia* were ____________.

(b) The active compounds were ____________ to the surfaces of implant materials.

(c) Bacteria were ____________ on treated and untreated surfaces.

5. Why was it important to test the effect of the active compounds on polymer and ceramic materials?

6. Explain why bacteria were cultured on surfaces treated with diterpenes and on untreated surfaces.

Indigenous medicine meets biomaterials development [cont.]

In practice, the surfaces of implanted biomedical devices are particularly prone to the build up of bacteria and this can happen weeks or even months after a device is put in. Bacterial infections are routinely controlled with antibiotics, but by the time an infection has been identified on an implant deep inside the body, the patient can already be seriously ill. 'If someone in their seventies has an artificial hip inserted and then they have an infection and have to go back to get the hip taken out and another put in, that's traumatic for someone who is already compromised health-wise,' says Hans. It also causes avoidable expense and wasted time for the healthcare system. The joining of the antibacterial diterpenes to the surface of the implants gets around these problems and will hopefully provide permanent protection, stopping the implant from ever harbouring infection. Preliminary results also suggest that mouse cells, unlike the bacteria, can attach and grow on some of the modified surfaces.

Multi-national healthcare companies have expressed interest in the technology and patents (legal ownership of an invention or process) have been taken out. However, Hans is clear that the team wants to ensure that the Aboriginal community also benefits from the exploitation of these novel compounds. 'We can learn so much from nature and traditional knowledge.'

Australian Microscopy and Microanalysis Research Facility 2009

artificial (*adj*) made; not natural
compromised (*adj*) weakened
harbour (*v*) to give a home to
prone to (*adj*) easily affected by
routinely (*adv*) usually
traumatic (*adj*) very worrying; upsetting

7. Describe where in the body these compounds may be used to prevent growth of bacteria.

8. An artificial hip is an implanted biomedical device. What are three other biomedical implants that would benefit from treatment with antibacterial diterpenes?

9. Explain why infections related to biomedical implants are of particular concern.

RATE MY UNDERSTANDING
Shade the face that shows your rating

8.4 Immunisation

Science as a human endeavour

FOUNDATION	STANDARD	ADVANCED

Two success stories

The principles of immunisation are the same for viruses and bacteria. The two diseases below are viral diseases.

Smallpox

In the 18th century, smallpox was a disease that killed many people. In 1788 an epidemic of smallpox spread through the part of England where Edward Jenner (1749–1823) lived. He was a doctor and he noticed that people who caught cowpox, did not get smallpox. Cowpox is a similar but milder disease to smallpox.

In 1796, Jenner carried out an experiment on one of his young patients, eight-year-old James Phipps. Jenner made two small cuts on James's arm and rubbed into them a small amount of pus from a cowpox infection. James developed the slight fever that is normal for a cowpox infection. He soon recovered. A few weeks later, Jenner again made small cuts on James's arm. This time pus from a smallpox infection was rubbed into them. James remained healthy—he was immune to smallpox. Vaccination as a treatment to prevent disease had begun.

Within a few years, vaccination against smallpox was widespread. However, nearly 200 years later, about 2 million people still died from the disease each year. In the 1960s the World Health Organization began a worldwide vaccination campaign to eradicate smallpox. In December 1979 the disease was declared extinct in nature.

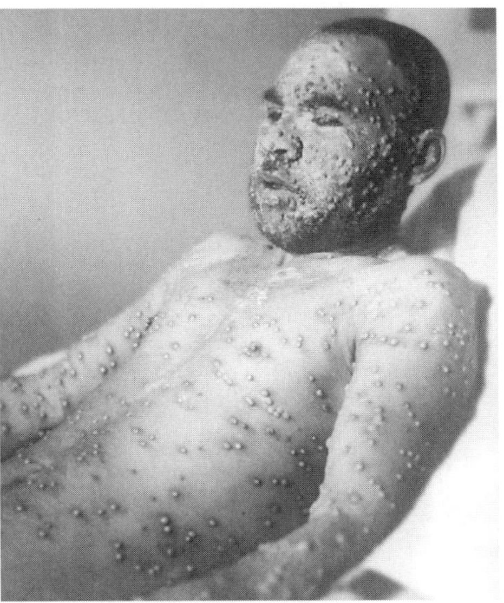

Figure 8.4.1 Smallpox was a viral disease that caused raised fluid-filled blisters. Estimates suggest that smallpox was responsible for between 300 and 500 million deaths during the 20th century.

Polio

Polio is a viral disease that causes muscle wasting, paralysis and sometimes death. In the early 1950s an American researcher, Doctor Jonas Salk (1914–1955), developed the first successful polio vaccine. To create the vaccine, Salk grew a live virus in the laboratory and then 'killed' it by adding the chemical formaldehyde. In 1954 there was nationwide testing of the vaccine on hundreds of thousands of schoolchildren. Unfortunately, one batch of the vaccine had not been made correctly and some children became ill, and a few died. Once the manufacturing process was improved, the vaccine successfully reduced the number of cases of polio.

In 1958, after 20 years of research, a Polish–American doctor, Albert Sabin (1906–1993), tested an alternative polio vaccine. Sabin's vaccine used weakened viruses and could be taken by mouth rather than by injection. It became the more popular vaccine because it was cheaper to manufacture and easier to administer (give).

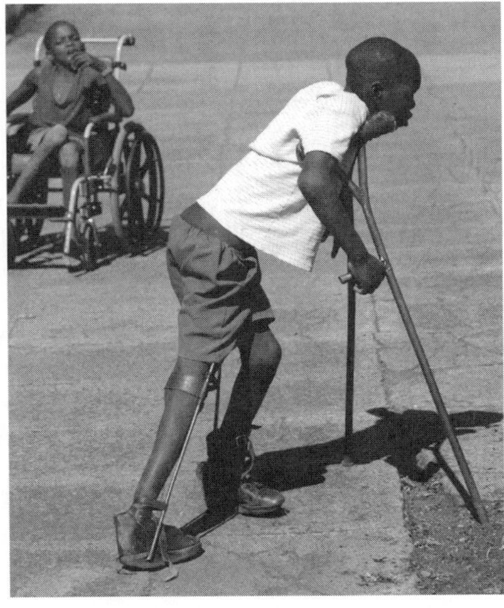

Figure 8.4.2 This boy has been crippled by polio

8.4 Immunisation

The two vaccines between them have eradicated polio from most countries in the world.

1 Calculate the number of years between the vaccination process being tested by Jenner and the eradication of smallpox.

____

2 Explain the term eradication.

____

③ Create a flow diagram of the process Jenner used to make James Phipps immune to smallpox.

④ What dangers do you think were associated with Jenner's experiment?

____

____

⑤ Look at Figure 8.4.2. The muscles of the boy's legs are wasted. What do you think 'wasting of muscles' means and what do you think causes it?

____

____

6 **(a)** Why did the first trials of the Salk polio vaccine cause problems?

____

____

(b) What was done to prevent these problems happening again?

____

____

⑦ The two vaccines for polio were made in different ways. Describe how each vaccine was made.

____

____

____

RATE MY UNDERSTANDING
Shade the face that shows your rating   

8.5 Viruses

Science understanding

FOUNDATION	STANDARD	ADVANCED

1 Choose one of the viruses in the box below to research. Highlight your chosen virus.

dengue	ebola	measles	mumps	rubella	zika virus

2 Research the selected virus and complete the boxes below. Information can be presented in written and/or visual form.

(a) The main countries affected:

(b) The symptoms of the virus:

(c) The effects of the virus (on people and around the world):

(d) The life cycle of the virus. Draw it in the space below.

(e) The people most at risk of infection:

____

____

____

____

____

(f) Treatments for the virus:

____

____

____

____

____

(g) Ways to reduce the spread of the virus:

____

____

____

____

____

RATE MY UNDERSTANDING
Shade the face that shows your rating

8.6 Swine flu pandemic

Science inquiry skills

FOUNDATION | **STANDARD** | ADVANCED

Processing & Analysing

On 24 April 2009, the United States government reported that a flu-like illness had reached the United States, with seven people confirmed as having influenza H1N1. Meanwhile, in Mexico there were 12 confirmed cases. Flu is not unusual, but this strain of flu was different. While the very young and very old are normally the people worst affected by flu, H1N1 was affecting young adults. The strain of flu was called swine flu, but it is not the same as the virus detected in pigs (swine). It was a new strain. An H1N1 virus is shown in Figure 8.6.1.

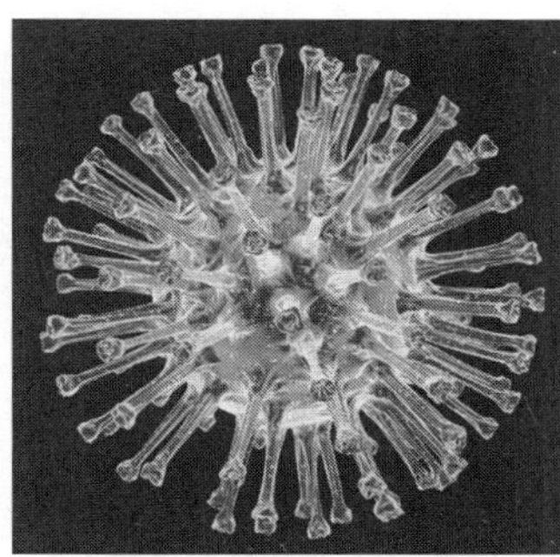

Figure 8.6.1 An H1N1 virus

confirmed (*adj*) checked as being true

strain (*n*) a variety or type

Australia recorded its first case in early May 2009. A month later the infection was declared a pandemic by the World Health Organization (WHO). A pandemic is an infectious disease that spreads rapidly over a wide area, and this infection was spreading very rapidly, as shown by the graph data in Figure 8.6.2.

The number of deaths from the disease continued to rise for the next few months (Figure 8.6.3).

On 10 August 2010, the WHO Director-General, Dr Margaret Chan, announced that the H1N1 influenza event had moved into the post-pandemic period and that although the disease would continue to be monitored there would no longer be weekly public updates.

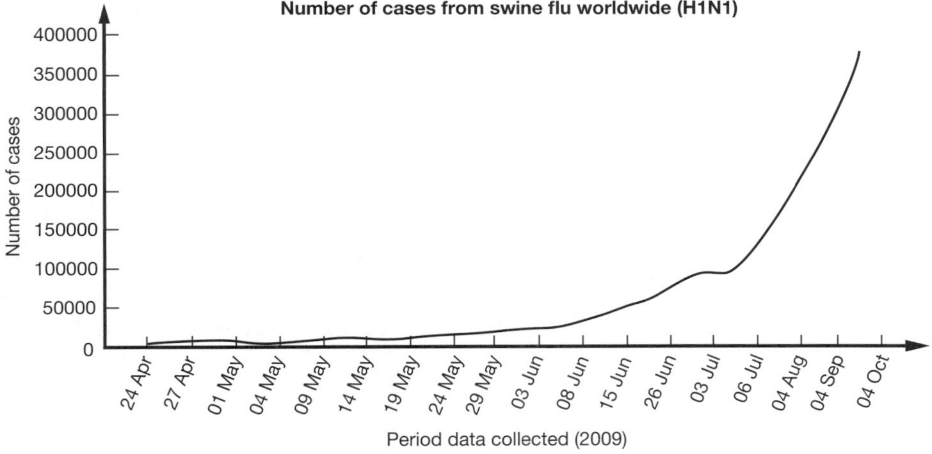

Figure 8.6.2 The number of cases of H1N1 virus increased slowly to begin with. The rate of infection then increased rapidly.

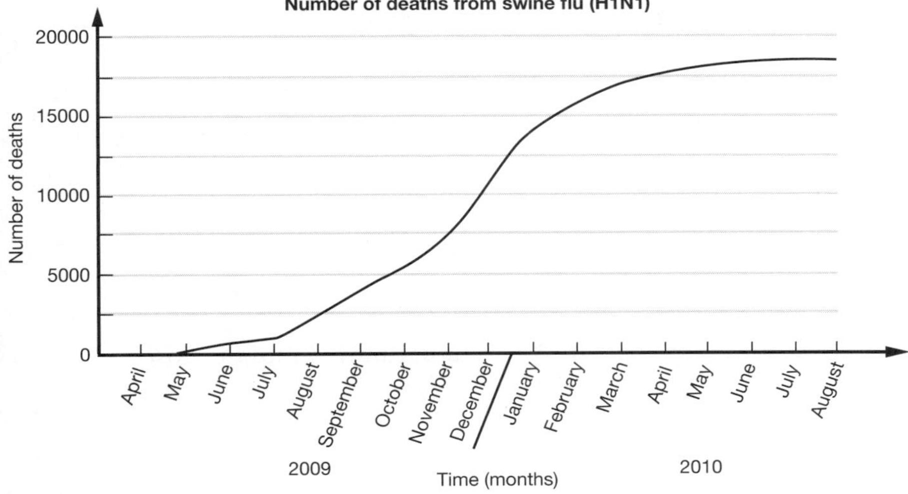

Figure 8.6.3 Number of deaths from H1N1 influenza

8.6 Swine flu pandemic

1 What are the differences between swine flu and normal flu?

2 Provide reasons for the WHO suggesting that people avoid unnecessary travel.

3 Explain why so many people became ill even though many had been vaccinated against flu viruses.

4 Explain why the spread of swine flu was called a pandemic.

(5) Refer to Figure 8.6.2 which shows the increase in the number of cases over the first few months of the H1N1 pandemic and complete the following sentences.

The number of people with swine flu, between 24 April and 15 June ______________

from 19 people to ______________ , an increase of ______________ percent.

The slope of the curve between 24 April to 15 June can be described as

______________ compared with the slope between 4 August to 4 October which can be

described as ______________ .

The difference in the slopes of the graph at these periods means that ______________

The differences in the slope at these periods can be explained by ______________

(6) Using the information in Figure 8.6.3, describe what happened to the number of deaths from H1N1 influenza from February 2010 to August 2010.

(7) Deduce why the announcement was made that 'the event has moved into the post-pandemic period'.

RATE MY UNDERSTANDING
Shade the face that shows your rating

8.7 What's in my food?

Science inquiry skills

FOUNDATION | **STANDARD** | ADVANCED

Processing & Analysing | Communicating

Food testing

A variety of foods were tested to find out the quantities of different nutrients present in them. The amounts of those nutrients were recorded in Table 8.7.1. The data provided is based on the same units of measurement: 100 g for solid foods and 100 mL for liquids.

Table 8.7.1 Nutrients in food

Food type	Nutrients in food								
	Energy (kJ)	Protein (g)	Fat (g)			Carbohydrate (g)		Calcium (mg)	Sodium (mg)
			Total	Saturated	Unsaturated	Total	Sugar		
Whole milk (100 mL)	276	3.2	3.9	2.4	1.5	4.8	4.8	120	44
Semi-skim milk (100 mL)	193	3.3	1.4	0.9	0.3	5.0	5.0	124	43
Tasty cheddar cheese (100 g)	1715	24.3	35.2	23.9	11.3	0.1	0.1	735	635
Pasta (100 g)	1489	12.0	1.8	0.5	1.3	70.4	<1.0	5	0
Rice (100 g)	550	1.9	3.3	0.5	2.8	23.3	0.1	0	3
Brown sugar (100 g)	1649	<1.0	0	0	0	96.8	96.8	0	21
Honey (100 g)	1416	0.3	0	0	0	83.1	77.4	0	15
Cashew nuts roasted with salt (100 g)	2520	15.9	47.2	9.6	37.6	27.5	5.4	0	240
Wholemeal bread (100 g)	989	9.1	3.9	0.5	3.4	37.6	3.0	0	400
Rump steak (100 g)	630	23	7	3	4	0	0	20	70
Minced beef (100 g)	880	20	25	9	16	0	0	20	70
Chicken breast (skinless) (100 g)	400	45	1	0.33	0.67	0	0	20	60
Chicken breast (skin on) (100 g)	545	45	5	1.6	3.4	0	0	20	60
Apple (100 g)	190	<1.0	0	0	0	12	0	20	0
Banana (100 g)	250	1	0	0	0	20	0	6	1
Broccoli (100 g)	100	2	<1	0	0	5	0	70	10
Green peas (100 g)	255	5.4	0.4	0	0	8	5.7	25	10
Lettuce (100 g)	30	0.7	0	0	0	3	0	20	10

8.7 What's in my food?

1. Use information from Table 8.7.1 on page 113 to respond to each statement. Circle whether it is TRUE, FALSE or you CANNOT TELL from the information provided.

Statement	True / False / Cannot tell
It is possible to compare the different foods because the same amount of food is analysed in each case.	True / False / Cannot tell
All liquids are measured in 100 gram units.	True / False / Cannot tell
The greatest range of nutrients was found in milk and cheese.	True / False / Cannot tell
The nutrient required for growth and repair is protein.	True / False / Cannot tell
Chicken breast and minced beef are the two foods with the highest amount of protein.	True / False / Cannot tell
Minced beef provides more energy than rump steak.	True / False / Cannot tell
Chicken with skin on has ten times more fat than chicken with skin off.	True / False / Cannot tell
Brown sugar has less total fat than honey.	True / False / Cannot tell
Apples, bananas and lettuce have no fat.	True / False / Cannot tell
There is more sodium in rice than pasta.	True / False / Cannot tell
It is a healthier choice to choose high sodium than high fat content foods.	True / False / Cannot tell
Foods with less than 10 grams of protein also have less than 200 KJ of energy.	True / False / Cannot tell

2. **(a)** Construct a bar or column graph to compare the amount of carbohydrate and protein provided by rice, bread and pasta. Include a scale and label the horizontal and vertical axes. Give the graph a heading.

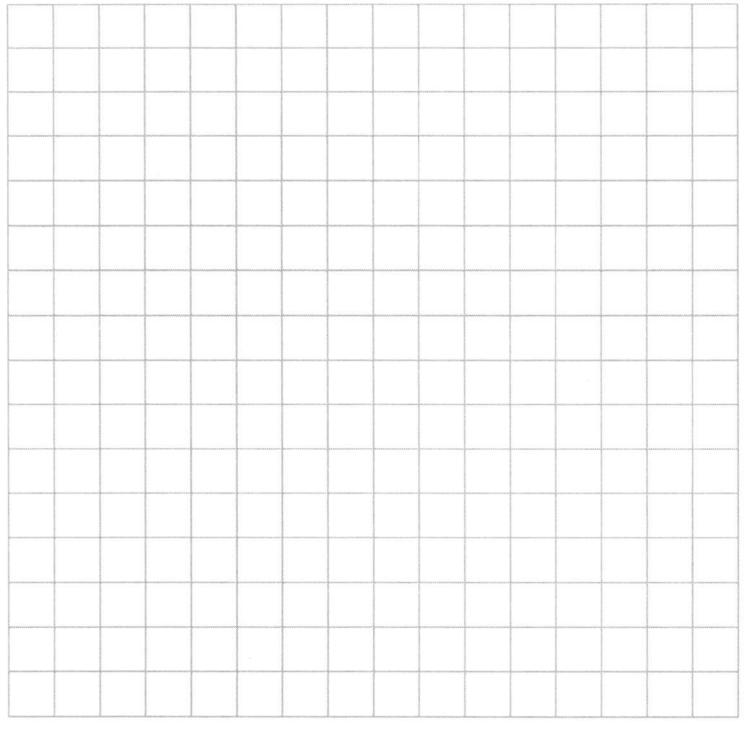

(b) State which of these foods–pasta, bread or rice–contributes most to a balanced diet.

---

(3) Explain why people trying to lose weight are encouraged to eat salads and lots of vegetables.

---

(4) Compare whole milk and semi-skim milk. As part of your comparison, calculate the percentage difference in the fat content.

---

(5) **(a)** If a person wanted to lose weight, then which three foods in Table 8.7.1 on page 113 should they avoid?

(b) Explain your answer.

---

(6) **(a)** If a person wanted to gain weight, then which foods would you encourage them to eat?

(b) Explain your answer.

---

(7) **(a)** Which foods listed should a person with high blood pressure avoid?

(b) Explain your answer.

RATE MY UNDERSTANDING
Shade the face that shows your rating   

8.8 Nutritional information labels

Science inquiry skills

FOUNDATION | **STANDARD** | ADVANCED

The following are nutritional panels from five snack bar packets.

NUTRITION INFORMATION #
Servings per Package: 6 Serving Size: 37.5 g (1 bar)

	Quantity per Serving	% Daily Intake* per Serving	Quantity per 100 g
ENERGY	530 kJ	6%	1420 kJ
PROTEIN	1.8 g	4%	4.8 g
FAT, TOTAL	1.1 g	2%	2.9 g
- saturated	0.3 g	1%	0.9 g
- trans	<0.1 g		<0.1 g
- polyunsaturated	0.4 g		1.0 g
- monounsaturated	0.4 g		1.0 g
CARBOHYDRATE	26.3 g	8%	70.1 g
- sugars	15.4 g	17%	41.0 g
DIETARY FIBRE	1.8 g	6%	4.7 g
SODIUM	60 mg	3%	166 mg

All specified values are averages
* Percentage Daily Intakes are based on the average adult diet of 8700 kJ. Your Daily Intakes may be higher or lower depending upon your energy needs.

Made in Australia from local and imported ingredients.

Figure 8.8.1 Fruit Breaks™

NUTRITION INFORMATION (AVERAGE)
Servings per package: 6
Average serving size: 32 g (1 BAR†)

	Quantity per Serving	% Daily Intake* per Serving	Quantity per 100 g
Energy	460 kJ	5%	1450 kJ
Protein	2.4 g	5%	7.5 g
Fat			
- Total	1.5 g	2%	4.8 g
- Saturated	0.4 g	2%	1.3 g
Carbohydrate			
- Total	19.8 g	6%	62.0 g
- Sugars	6.9 g	8%	21.6 g
Dietary Fibre	3.7 g	12%	11.5 g
Sodium	51 mg	2%	160 mg
Potassium	80 mg		250 mg

† Bar weight is approximate and is only to be used as a guide. If you have any specific dietary requirements please weigh your serving.

* % Daily Intakes are based on the average adult diet of 8700 kJ. Your daily intakes may be higher or lower depending upon your energy needs.

CONTAINS 49% WHOLE GRAINS

Figure 8.8.2 Trail Bars®

Nutrition Information
(AVERAGE)
Servings per package - 6
Average serving size - 22 g (1 BAR†)

	quantity per serving	% daily intake ▲per serving	quantity per 100 g
ENERGY	370 kJ	4%	1700 kJ
PROTEIN	0.7 g	1%	3.1 g
FAT - TOTAL	2.0 g	3%	9.2 g
- SATURATED	0.5 g	2%	2.3 g
CARBOHYDRATE	16.9 g	5%	76.9 g
- SUGARS	7.5 g	8%	34.3 g
DIETARY FIBRE	0.1 g	0.4%	0.5 g
SODIUM	72 mg	3%	325 mg
POTASSIUM	14 mg		65 mg

†Bar weight is approximate and is only to be used as a guide. If you have any specific dietary requirements please weigh your serving.

▲% Daily Intakes are based on the average adult diet of 8700 kJ. Your daily intakes may be higher or lower depending upon your energy needs.

Figure 8.8.3 LCMs®

Nutrition Information
Servings per package: 6
Average serving size: 35 g (1 bar)

	Avg. Quantity per serving	Avg. Quantity per 100 g
Energy	713 kJ	2040 kJ
Protein	5.3 g	15.1 g
Fat, total	10.7 g	30.6 g
- saturated	2.5 g	7.2 g
Carbohydrate, total	13.8 g	39.3 g
- sugars	8.6 g	24.5 g
Dietary Fibre	2.5 g	7.1 g
Sodium	8 mg	22 mg

All specified values are averages

TM Trademark

Figure 8.8.4 Natural Nut Bars™

NUTRITION INFORMATION
Servings Per Package: 8
Serving Size: 31.25 g

	Avg. Quantity Per Serving	Avg. Quantity Per 100 g
Energy	502 J	1610 kJ
Protein	1.9 g	6 g
Fat, total	3.0 g	9.4 g
- saturated	LESS THAN 1 g	2.4 g
Carbohydrate	20.4 g	65.2 g
- sugars	5.8 g	18.5 g
Dietary Fibre	2.1 g	6.8 g
Sodium	LESS THAN 5 mg	9 mg

Figure 8.8.5 Three Fruits Muesli Bars

1 Use the information on the nutrition panels to complete the table below. 100 grams of each bar contains:

Snack bar type	Energy (kJ)	Protein (g)	Total fat (g)	Total carbohydrate (g)	Sugar (g)	Fibre (g)	Sodium (mg)
Fruit Breaks							
Trail Bars							
LCMs							
Natural Nut Bars							
Three Fruits Muesli Bars							

2 Construct column graphs to compare the amount of each nutrient per 100 g of the snack bars.

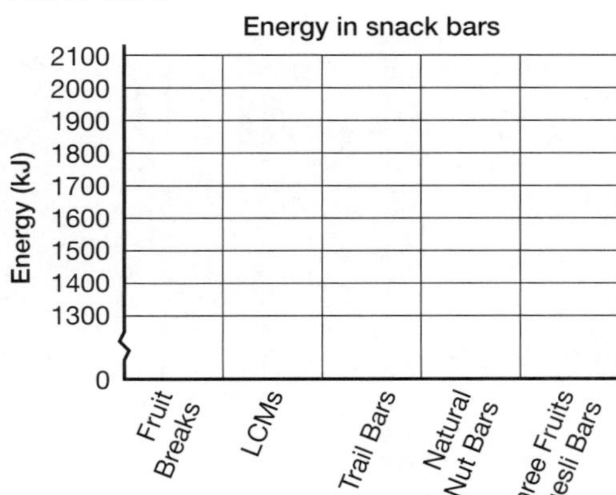

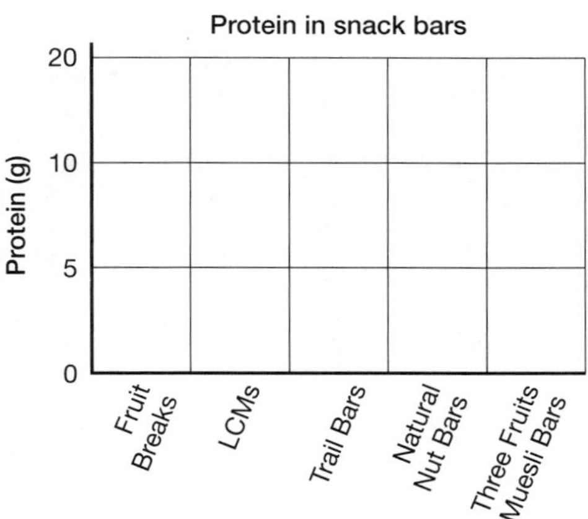

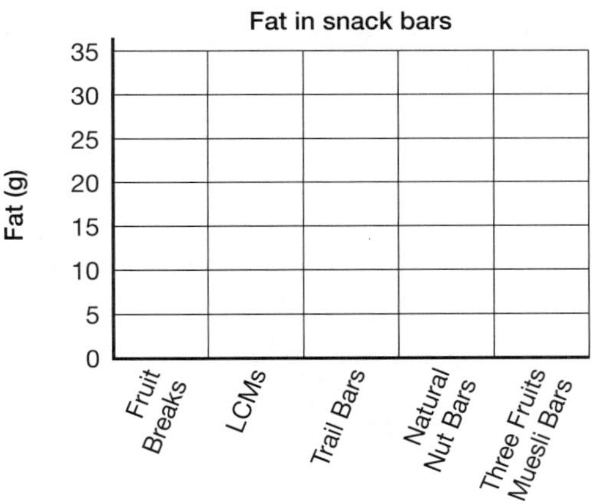

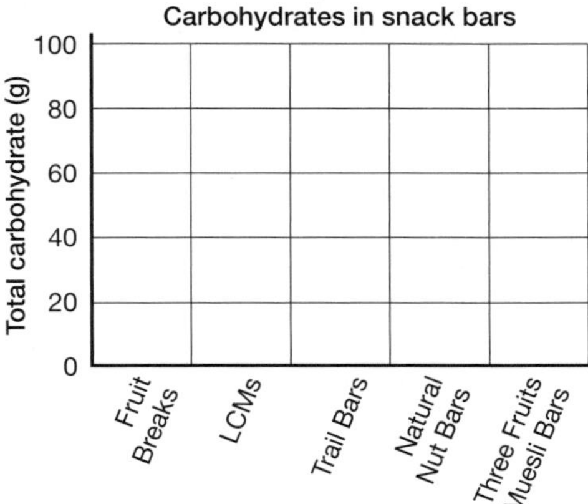

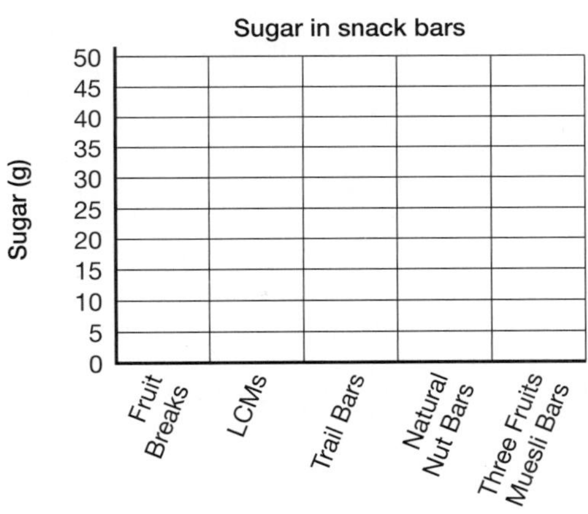

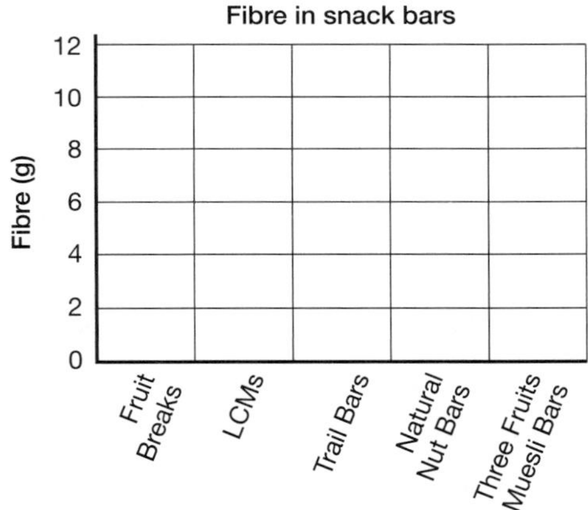

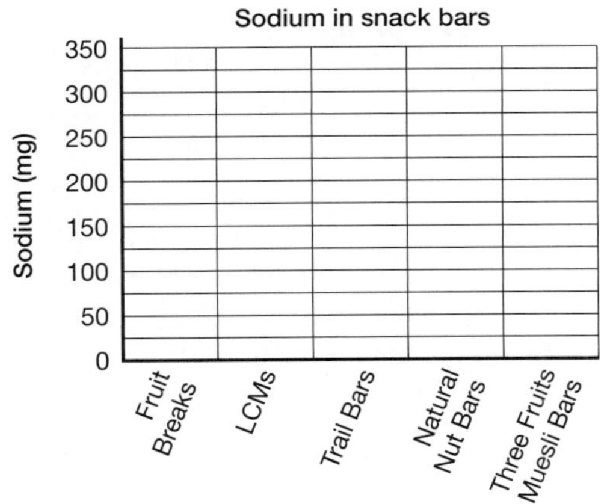

3 Analyse the graphs then list the snack bars in order of greatest to least in amount (per 100 g) of each nutrient listed.

Snack bar nutrition rank	Nutrients						
	Energy	Protein	Fat	Total carbohydrate	Sugar	Fibre	Sodium
5 greatest amount							
4							
3							
2							
1 least amount							

4 **(a)** Which snack bars best support healthy eating?

____

(b) Justify your answer.

____

____

5 **(a)** What changes do you think would improve the snack bar identified in question 4 to make it healthier to eat?

____

____

(b) Explain why these changes would help.

____

____

6 **(a)** Name the snack bar that is least supportive of healthy eating.

____

(b) Explain your answer.

____

____

RATE MY UNDERSTANDING
Shade the face that shows your rating   

8.9 Literacy review

Science understanding

FOUNDATION	STANDARD	ADVANCED

1 Use words from the box to complete the sentences below. Words may be used more than once.

antibiotics	chronic diseases	contagious	diabetes
disease	infectious	immunity	lymphocytes
neutrophils	nutrition	over nutrition	parasites
pathogens	quarantine	saliva	skin
stomach acid	under nutrition	vaccines	

(a) A ____________________ is anything that causes your body to stop working correctly.

(b) Bacteria and viruses are two types of ____________________ that cause ____________________ .

(c) Diseases passed easily from one person to another are ____________________ diseases.

The most easily passed on are known as ____________________ .

(d) People who are contagious are put in ____________________ . Separating them from others helps control the spread of the disease.

(e) Bacterial diseases are treated with ____________________ .

(f) ____________________ , ____________________ and ____________________ are all part of the body's first line of defence.

(g) ____________________ in the blood are an important part of the body's second line of defence.

(h) When your body is infected with a pathogen, your immune system makes antibodies,

which give you ____________________ the next time you meet the same pathogen.

Antibodies are made by ____________________ .

(i) ____________________ are often given by injection. They cause your body to make antibodies.

(j) Some diseases are caused by microscopic ____________________ such as Plasmodium.

(k) ____________________ is the food necessary for health and growth. Too much food

leads to ____________________ . Too little food or food lacking nutrients leads to

____________________ .

(l) ____________________ are diseases that last for a long time.

(m) ____________________ is a complex, chronic disease caused by a lack of insulin or the inability of the body to use insulin.

RATE MY UNDERSTANDING
Shade the face that shows your rating

8.10 Thinking about my learning

Imagine you are the world renowned expert in infectious diseases, Professor Path Ogen. You have been asked to talk to a class of students about your work with infectious diseases and related health issues. You have been asked to speak for 15 to 20 minutes with another 10 minutes devoted to questions.

Your task is to prepare four A3-size prompt sheets outlining your subject material that you can refer to while giving your talk. Your information should be in point form. The areas you need to cover are as follows:

- The diseases we need to protect our community against
- The ways our body defends itself against disease
- The medicines we have available to combat infectious diseases
- The role of vaccination in protecting against infectious diseases

You also need to prepare three or four questions, and their answers, on another card that might arise from your talk about infectious diseases.

Infectious diseases and health issues question card
Question 1
Question 2
Question 3
Question 4

CHAPTER 9

Ecosystems

9.1 Knowledge preview

Science understanding

FOUNDATION | **STANDARD** | ADVANCED

All plants and animals have adaptations that enable them to live in their environments. Some of these environments are very extreme, such as deserts, the deep sea or cold continents like Antarctica.

Figure 9.1.1 Polar bear

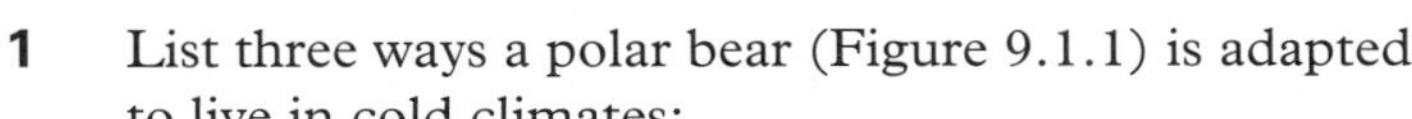

1 List three ways a polar bear (Figure 9.1.1) is adapted to live in cold climates:

(a) ____________________

(b) ____________________

(c) ____________________

2 (a) Figure 9.1.2 below shows a simple Antarctic Ocean food chain.

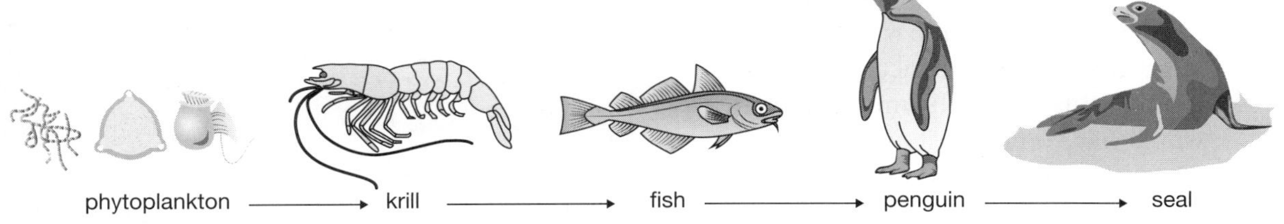

Figure 9.1.2

(i) What is a food chain? ____________________

____

(ii) Identify the 'producer' in this food chain. ____________________

(iii) Explain why you selected this marine animal as the producer.

____

____

____

(b) Joining food chains together produces a food web. Redraw the Antarctic Ocean food chain in Figure 9.1.2 to create a food web. Refer to the following facts about Antarctic animals for assistance. Use the space provided on the next page.

- elephant seal eat fish
- birds eat fish
- fish are eaten by penguins
- birds are eaten by leopard seals
- baleen whales eat krill
- smaller toothed whales eat penguins
- smaller toothed whales eat leopard seals
- elephant seals are eaten by smaller toothed whales
- baleen whales are eaten by smaller toothed whales

9.1 Knowledge preview

Construct your food web here.

3 Changes to a food web will not only affect individual organisms, but it will have a flow-on effect for the whole food web. Look carefully at the food web of a freshwater stream (Figure 9.1.3). Assume that pollution from a nearby stream has killed off large numbers of the bacteria in this food web. Evaluate the impact this would have on the rest of the organisms in the food web.

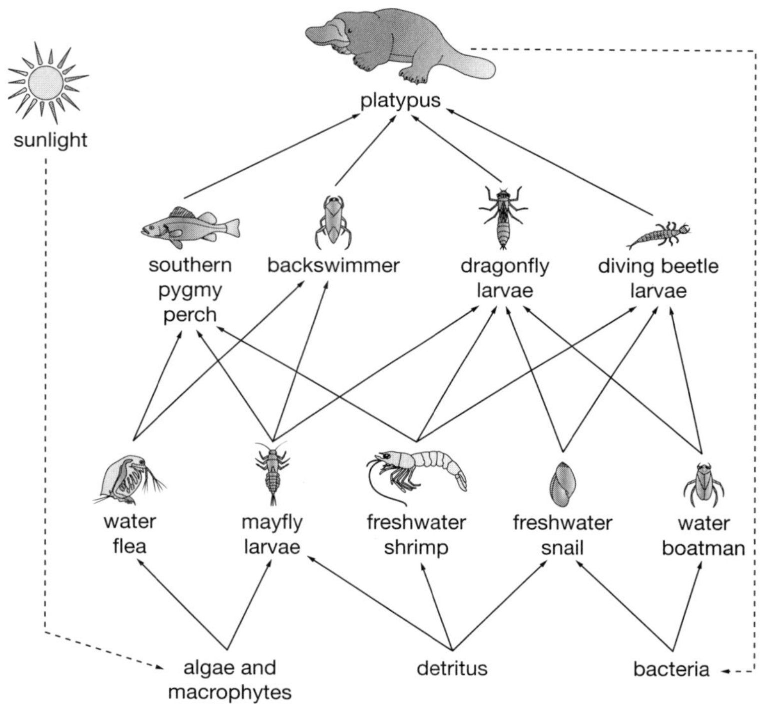

Figure 9.1.3

9.2 Germinating seeds

Science as a human endeavour

FOUNDATION | **STANDARD** | ADVANCED

Seeds of some plant species (such as wattles) germinate better after bushfires. Germination is the process of seeds growing and sprouting. For most seeds, germination just involves coming into contact with water. These seeds absorb water, swell up and begin growing. Some Australian plants, especially wattles, have seeds with tough waterproof coats. Scientists discovered that if this coating is split, water can enter. Experiments showed that soaking wattle seeds in boiling water for a few minutes could split the seed coat. So, for a long time, ecologists thought it was only heat that causes so many wattles to germinate after a bushfire. They also thought this should be true for other plant species as well.

flora (*n*) plants

horticulture (*n*) the practice of managing and growing a garden of flowers, fruits, vegetables

nursery (*n*) a place where plants are grown to be sold

sprout (*v*) to begin growing, to put out shoots

Before 1990, many species that the horticultural industry and researchers had been trying to grow would not germinate very well. For these 'problem species', heat treatment did not work, and the researchers had no idea why. For example, in Western Australia, about 20% of the native species were difficult to germinate. This meant that these species could not be:

- grown in plant nurseries and sold to the public
- widely used in repairing natural ecosystems damaged, for example, by mining.

Anyone who wanted to use these species had to obtain their seeds from the wild. This put pressure on the natural ecosystems.

1 Define the term germination.

____

____

2 Explain why, before 1990, Australian scientists thought it was probably heat that was causing wattles and other plants to germinate following a fire.

____

____

____

3 Explain why Western Australian scientists were uncertain about what to do with 'problem species'.

____

____

____

4 What are three benefits of improving the germination of 'problem species'?

____

____

____

9.2 Germinating seeds

In 1990, two South African scientists discovered that smoke alone, rather than heat, improved the germination of some South African plants. The climate, flora and bushfire history of South Africa is similar to that of Western Australia. So scientists at Kings Park and Botanic Garden in Western Australia decided to try smoke on Western Australian plants. They discovered a dramatic response of many species to smoke.

This caused more research around Australia. Over 400 species have been shown to germinate better after smoke treatment. Because of this research, smoke is now widely used in nursery production, bushland management and mine-site restoration in Australia. It is even being used by home gardeners and farmers who are interested in native plants.

aerosol

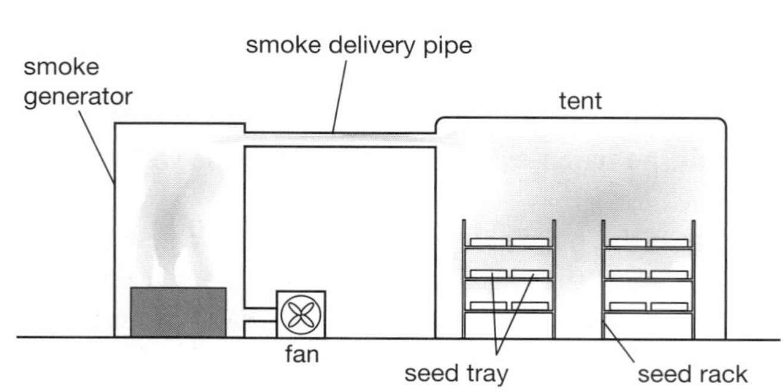

smoke water

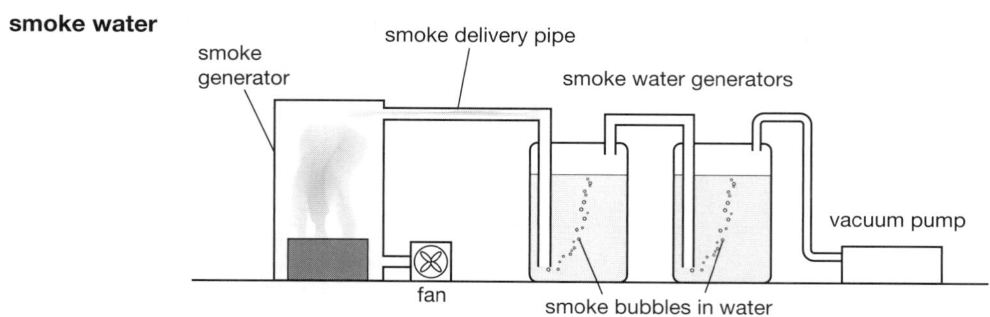

Figure 9.2.1 Smoke can be applied to the seeds by air (called 'aerosol') or water (called 'smoke water'). These methods can be applied to trays of seeds in soil, bushland soil or directly to seeds. The method used depends on what suits the situation.

5 Explain why scientists at Kings Park and Botanic Garden decided to try smoke to improve germination of Western Australian native plants.

____

____

____

6 Describe the two methods by which smoke is applied to the seeds.

____

____

____

RATE MY UNDERSTANDING Shade the face that shows your rating	☺		

9.3 Biotic and abiotic factors

Science understanding

FOUNDATION	**STANDARD**	ADVANCED

1 The following table lists some physical abiotic factors in an ecosystem. Complete the table by explaining at least one way each factor can affect living organisms.

Effects of abiotic factors on living organisms	
Physical factor (abiotic factor)	**One way in which this factor can affect living organisms**
water	
temperature	
fire	
light	
soil minerals	
oxygen level	

2 Complete the following table. Describe the meaning, or recall an example of each of the biotic factors listed.

Biotic factors		
Biotic factor	**Meaning/definition**	**Example**
competition		baby birds in the nest trying to get food from the parent
predation	one organism (predator) killing and eating another (prey)	
mutualism		the flagellates in termite guts
commensalism	organisms living together where one benefits and the other is unharmed	
parasitism		tapeworm and human
decomposers		fungus decaying a dead tree

RATE MY UNDERSTANDING
Shade the face that shows your rating

9.4 Biotic and abiotic factors in the ocean depths

Science understanding

FOUNDATION | **STANDARD** | ADVANCED

The diagrams in Figure 9.4.1 show six fish that live in the ocean depths. Study the illustrations and answer the questions that follow.

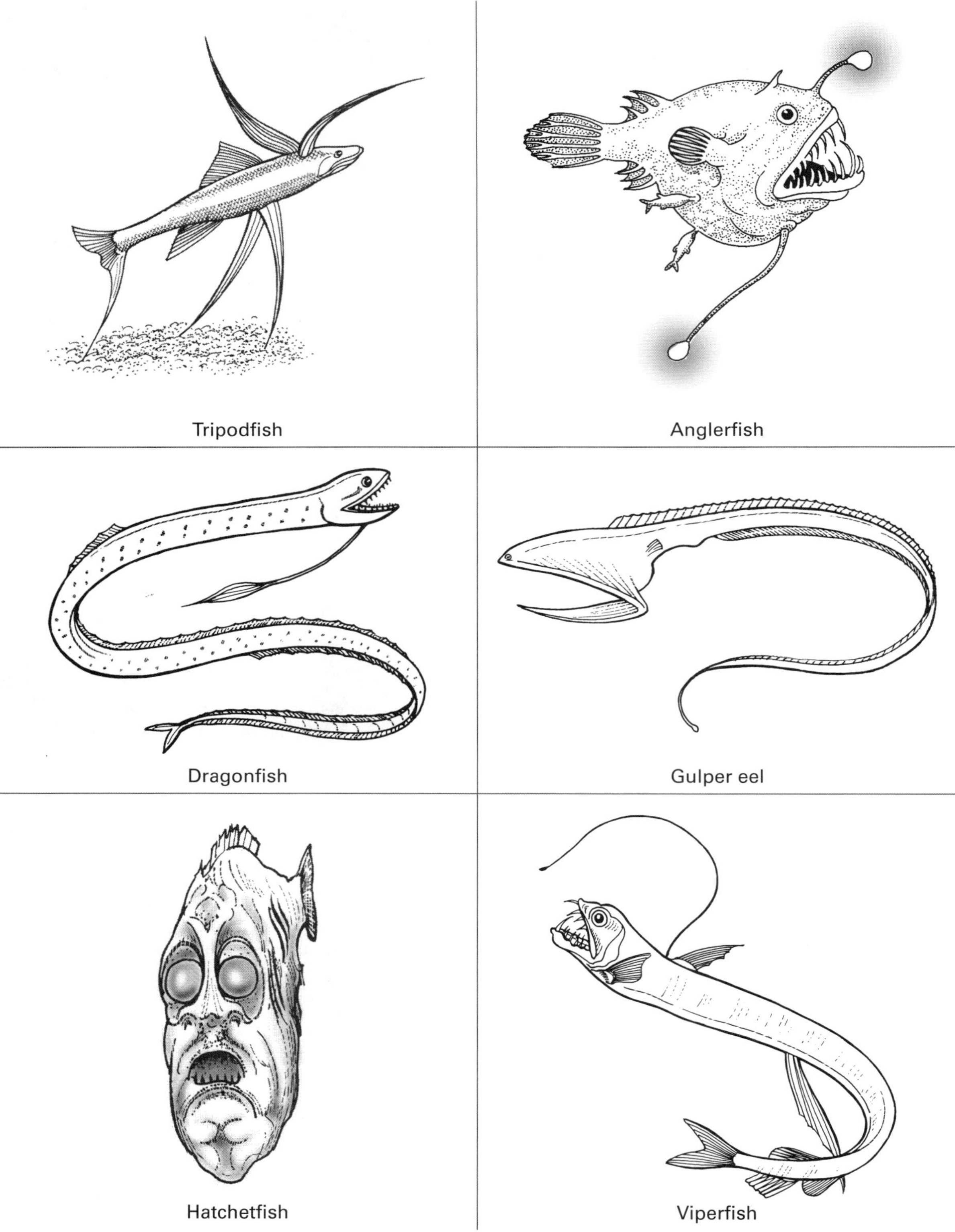

Figure 9.4.1

9.4 Biotic and abiotic factors in the ocean depths

lure (*n*) something used to attract attention

predator (*n*) animal that hunts and kills another animal for food

prey (*n*) an animal that is hunted and eaten by another animal

1. The fish in Figure 9.4.1 are all predators, but most live in places that are completely dark. Some of these predators have 'lures' that make light and are waved around to attract curious fish. Identify the abiotic factor mentioned above and explain how it affects survival of the predators.

____

____

2. The deep ocean is low in oxygen, which means individuals have little energy available from respiration to keep them active. Populations are very low for most species because there is limited food available. The tripod fish has poor vision and stands on the bottom of the ocean on three stiff fins waiting for its prey to bump into it. Identify an abiotic factor in the environment of a tripod fish and explain how this factor affects the population size of the fish.

____

____

____

HINT

The low level of oxygen is a key abiotic factor for this fish. How does it affect the population of the tripod fish?

3. The hatchetfish is a predator that has excellent vision. Cells on its belly can make blue light. This colour light matches the faint blue glow from shallower water in areas where some light penetrates the water. How do you think the production of blue light (a biotic factor) affects the hatchetfish in the deep sea environment?

HINT

What colour does the hatchet fish produce? What colour is the surrounding sea? How are these factors helpful for survival?

____

____

____

____

____

4. Many of the fish in this environment have extremely long and sharp teeth. Explain how these teeth can be considered an adaptation for life in the ocean depths.

____

____

5. The gulper eel has a huge mouth. It normally eats small, slow prey such as prawns and small fish. It can unhinge its jaw to make its mouth open extremely wide, and it can eat a fish as large as itself. Explain how biotic and abiotic factors interact to affect the survival of the eel.

____

____

____

RATE MY UNDERSTANDING
Shade the face that shows your rating

9.5 Food webs and energy

Science understanding

FOUNDATION | **STANDARD** | ADVANCED

1. Use Figure 9.5.1 and Table 9.5.1 to help you construct a food web. Write the names of the organisms and connect them with arrows to show the direction in which food material (matter) and energy passes. Use the space on the following page to draw your food web.

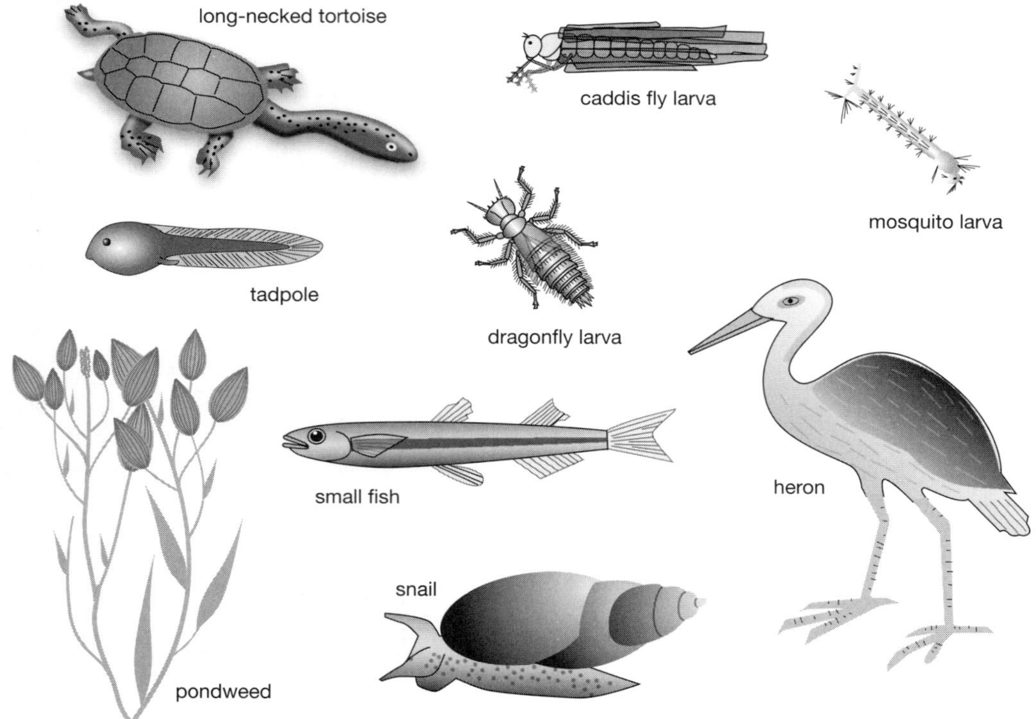

Figure 9.5.1

Table 9.5.1

Organism	Food
long-necked tortoise	snails, mosquito larvae, dragonfly larvae, tadpoles, caddis fly larvae, small fish
caddis fly larva	pondweed
mosquito larva	pondweed
tadpole	pondweed
dragonfly larva	caddis fly larvae
small fish	dragonfly larvae, mosquito larvae
heron	small fish, tadpoles
snail	pondweed
pondweed	photosynthesis

9.5 Food webs and energy

Construct your food web here.

(2) Where and how does energy enter this food web?

(3) Choose any consumer in the food web and explain how it obtains energy and what happens to that energy.

4 From the web, construct one food chain with five levels in it and another with three levels.

(5) Select one of your food chains in question 4. Describe what happens to the amount of energy in the matter passing along the chain.

RATE MY UNDERSTANDING
Shade the face that shows your rating

9.6 Biotic factors and population changes

Science inquiry skills

FOUNDATION | **STANDARD** | ADVANCED

Processing & Analysing **Communicating**

Consider the following food web in a pond.

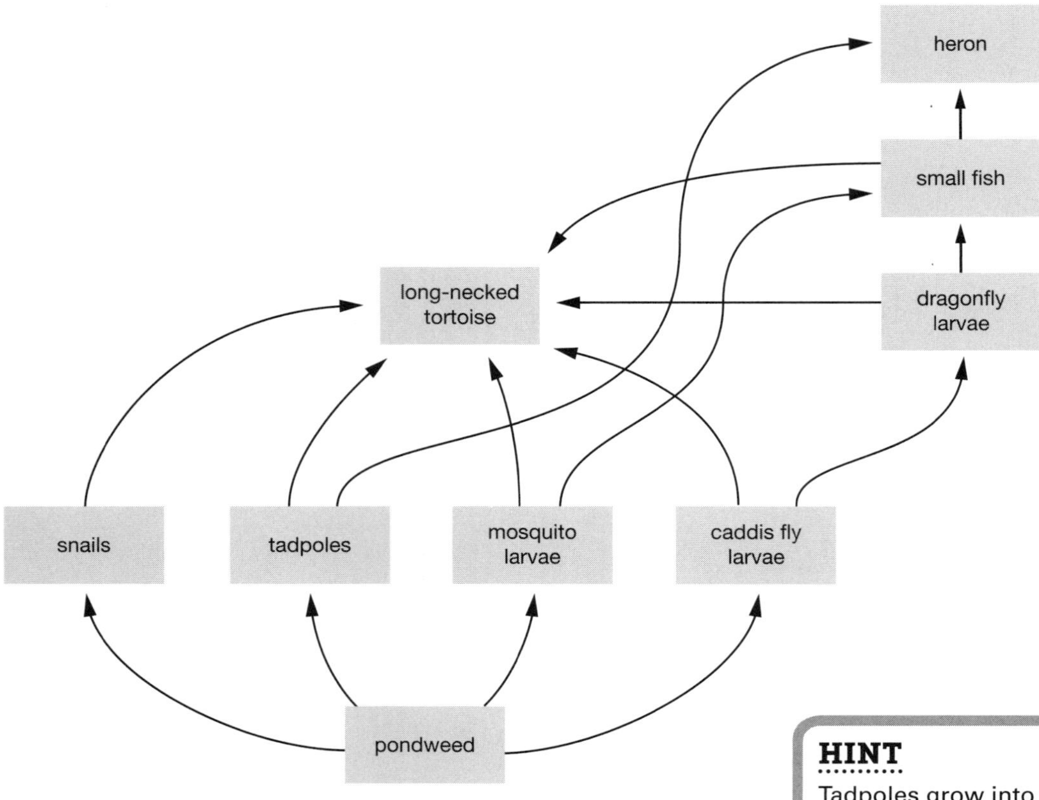

Figure 9.6.1 Food web in a pond

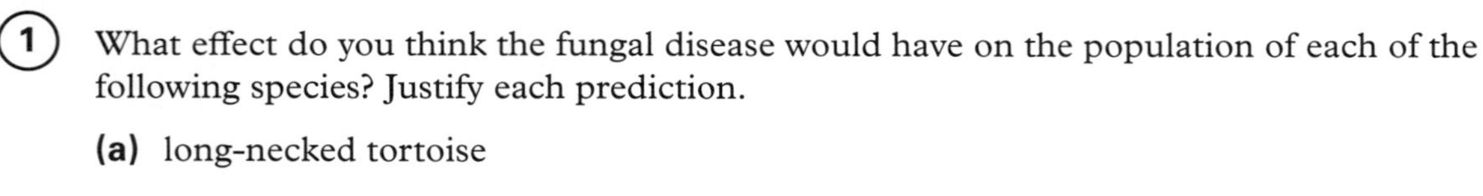

HINT

Tadpoles grow into frogs.

Imagine a new fungal disease enters the ecosystem and seriously reduces the number of frogs. Base your answers to the following questions on the food web in Figure 9.6.1.

1. What effect do you think the fungal disease would have on the population of each of the following species? Justify each prediction.

   (a) long-necked tortoise

   (b) snail

(c) pondweed

2. Assess whether the fungal disease would have more effect on the long-necked tortoise or the heron. Justify your prediction.

3. In a normal pond there is no fungal disease. The following graph shows the changes in the populations of tadpoles and long-necked tortoises in the pond over several years.

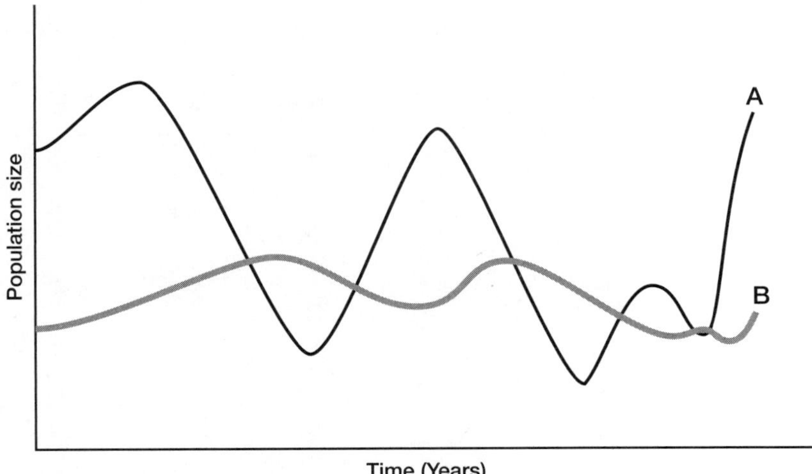

Figure 9.6.2 Population sizes of two species, A and B

(a) Identify the species represented by line B.

(b) Justify your answer to part (a).

9.7 Recovery plans for threatened species

Science as a human endeavour

FOUNDATION	STANDARD	ADVANCED

commercial (*adj*) to do with business

conservation status (*n*) the chance of a species surviving in the future

overhead (*adv*) over, above

recovery (*n*) return to good health

wipe out (*v*) to completely destroy

The Australian federal, state and territory governments have laws that protect threatened species. The Australian Government requires a document called a recovery plan for each species that is recognised as threatened. In the states and territories, there are similar laws about recovery plans for threatened species. A recovery plan for a species is a document that:

- describes the current conservation status
- summarises current biological and ecological knowledge of the species
- gives details of past and current management actions undertaken
- details a program for the next 5 years to help the recovery of the species.

The Wollemi Pine

The Wollemi pine shown in Figure 9.7.1 is of international importance because its ancestors date back to the time of the dinosaurs. The discovery of the Wollemi pine was a surprise because it was thought to be extinct. But about 100 adult trees and about 300 young trees were found in one area in the Blue Mountains north-west of Sydney. No other populations of the species have been discovered even after intensive searches. Consequently, the New South Wales Government acted fast to try to save the species. A fire in the area could have wiped out the entire world population. The Wollemi recovery plan has six objectives (Figure 9.7.1).

The Wollemi pine recovery plan objectives are to:

1. ensure long-term protection against processes that threaten the wild plants and their habitat
2. understand the biology of the Wollemi pine in its environment (its ecology) to make informed decisions to protect it
3. breed more of the species in other places to assist in its long-term survival
4. educate the public about the plant and its fragile environment
5. support commercial production of Wollemi pines for sale
6. coordinate the recovery of the Wollemi pine.

Figure 9.7.1 The Wollemi pine

9.7 Recovery plans for threatened species

1. Objective 5 recommends commercial production of Wollemi pines. The decision was made to propagate as many plants as possible and sell them to nurseries throughout Australia. Suggest how this protects the wild population.

2. The recovery plan (objective 5) considers the possible commercial value of the Wollemi pine. Justify the inclusion of this objective in the recovery plan.

3. Consider objective 3. Suggest how this could be of benefit to the conservation of the species.

4. Justify why understanding the ecology of the Wollemi pine (or any species) is essential to its recovery (objective 2).

5. Wollemi National Park has been established in the area around the Wollemi pine population. Only a few researchers are allowed into the area where the trees are found. All must disinfect their feet before entering the area. Aircraft are forbidden from flying overhead. Justify why these rules are included in the recovery plan.

9.7 Recovery plans for threatened species

6. Suggest three economic benefits of conserving the Wollemi pine.

1 ______________________________

____

2 ______________________________

____

3 ______________________________

____

7. Maintaining the ban on visiting the Wollemi pine in the Wollemi National Park is costly. Explain why a state government would be willing to spend this money.

____

____

____

____

____

8. Do you think it is important to conserve species like the Wollemi pine? Provide reasons to support your opinion.

____

____

____

____

____

____

____

____

____

____

RATE MY UNDERSTANDING
Shade the face that shows your rating   

9.8 Advantages of biological control

Science understanding

FOUNDATION	STANDARD	**ADVANCED**

Biological control is the use of a natural enemy (such as a predator) of a pest to control populations of that pest. This method of control, when done correctly, has great advantages because it does less damage to the environment than chemical pesticides. An example of insect populations being controlled biologically is shown in Figure 9.8.1.

Figure 9.8.1 A parasitic wasp named *Trioxys* lays an egg in an aphid. The egg hatches into a larva that eats the insides of the aphid before changing into an adult wasp

Some pesticides can accumulate in food webs, leading to sickness and death in higher order consumers. Biological control avoids this accumulation of poisons in food webs.

In addition, chemical pesticides will usually kill many other useful organisms, such as wasps and praying mantises, that may be natural predators of the pests. So, pesticides may actually make a problem worse. Pesticides also kill other useful organisms such as bees, which may be helping to pollinate flowers. Biological control has a great advantage over pesticides because it targets only the pest.

Over time, pests can become resistant to the chemical pesticide. This means that some populations of the pest are able to survive the poison. When this occurs, a stronger concentration of the pesticide or a different pesticide needs to be used to control pest populations. In biological control the pest never becomes resistant to the biological control agent.

Another advantage of biological control is that it can become permanently established. It does not have to be regularly applied like pesticides. Once the control agent is released, it will keep reproducing and so become a permanent part of the ecosystem. Therefore control over the pest is ongoing.

There are some disadvantages of biological control. One problem is that many pests come from overseas, so their natural enemy has to be imported. However, such control agents may not be suited to the climate in Australia. When introducing a biological control agent, care must be taken to ensure it only kills the pest and not other organisms that are found naturally in the environment. These checks are difficult to do and take time. Pesticides work quickly and do solve the problem, at least for a short time.

1 Define the term 'biological control'.

____

____

9.8 Advantages of biological control

2 Explain why pest resistance is an environmental problem.

______

______

______

______

3 Identify the advantages (plus-P), the disadvantages (minus-M) and interesting facts (I) of the use of biological methods of pest control. Use the PMI chart provided to write your answer.

Biological methods of pest control		
Plus	Minus	Interesting

4 Identify the advantages (plus-P), the disadvantages (minus-M) and interesting facts (I) of the use of chemical methods of pest control. Use the PMI chart provided to write your answer.

Chemical methods of pest control		
Plus	Minus	Interesting

5 Suggest why it could be a good idea in your home garden to carefully observe any insect pests on the plants to help you decide if you should spray them with pesticides.

______

______

______

6 Suggest why farmers still use pesticides even though they can cause environmental problems.

______

______

______

RATE MY UNDERSTANDING Shade the face that shows your rating			

9.9 Dung beetles

Science as a human endeavour

FOUNDATION	STANDARD	ADVANCED

Dung beetles are insects whose young eat animal dung (faeces). There are many different species of dung beetles. The adult beetles bury the dung under the soil and lay their eggs in it. This process means that dung is removed from the soil surface and takes away a food source for bushflies. Bushflies also lay their eggs in animal dung. Bushflies cause disease in farm animals such as sheep.

Dung beetles are widely used throughout Australia as biological control agents for bushflies. You can see beetles rolling dung in Figure 9.9.1.

Figure 9.9.1 Beetles rolling dung

Some agricultural scientists thought that there may be another benefit of dung beetles. The scientists thought that the beetles could be making the soil more fertile when they buried the dung. They knew that plant nutrients such as nitrogen are found in animal dung. The scientists predicted that these nutrients could make the soil better for growing pasture plants that feed farm animals. This was their hypothesis. An experiment was conducted in a cattle paddock in Merton, Victoria, to test their hypothesis.

The experiment used a grid of 54 rectangular 'plots' about 50 cm by 50 cm in a paddock for cattle grazing. Each plot had three replicates and was covered by a metal mesh cage. You can see them in Figure 9.9.2.

Each plot was given one of three treatments: 'dung and beetles', 'dung only' or 'control' (no dung or beetles). There were 54 plots in total, 18 for each treatment.

Figure 9.9.2

9.9 Dung beetles

Early results

The experiment is still continuing, but some results have been collected. After a period of time the plants in each of the 54 plots were cut off at ground level and placed in bags. The plants were then dried and weighed. The data collected is shown in Table 9.9.1.

Table 9.9.1

Dry weight of plants per plot	
Treatment	**Average dry weight per plot (g)**
control	90.96
dung	122.95
dung and beetles	148.24

1 What were the scientists testing in their experiment?

2 Identify the two experimental variables, dependent and independent, tested in this experiment.

3 Explain why control plots were used in this experiment.

4 Explain what is meant by a replicate.

5 How many replicates were used in this experiment?

6 Evaluate whether the results support the scientists' hypothesis.

RATE MY UNDERSTANDING
Shade the face that shows your rating

9.10 Literacy review

Science understanding

FOUNDATION	STANDARD	ADVANCED

1 Use the words in the box to complete the statements about ecosystems. You do not need to use all of the words, and words can be used more than once.

adaptation	biodiversity	conservation	ecology
ecosystems	introduced	living	organisms
overcropping	photosynthesis	rainforest	sustainable

(a) __________________ is the study of how organisms interact with each other and with their environment.

(b) Places in which __________________ and their physical surroundings form a balanced environment that is different from others nearby are called __________________ .

(c) The process whereby plants produce energy from sunlight is called __________________ .

(d) Humans are damaging the environment through practices such as __________________ which is killing more animals than the population can replace by its normal breeding cycle.

(e) Humans are also responsible for bringing animals to Australia from other countries. These are referred to as __________________ species.

(f) An __________________ is a feature that helps an organism survive and reproduce in its environment.

(g) Humans are trying to manage and protect __________________ so they can continue to exist. This is known as __________________ and is important if ecosystems are to be __________________ (able to continue to function on their own).

2 Match the definition in the first column with the correct term in the second column by connecting them with a line.

organisms trying to use the same resources
all the factors in an organism's surroundings that affect it
produced when lots of food chains join together
non-living factors
to cause damage to the place that an organism depends on for survival
species that are vulnerable to endangerment in the near future
to control a pest by introducing a natural predator
organisms that break down dead bodies and waste, and recycle material

environment
biological control
threatened species
food web
competition
abiotic
decomposer
habitat destruction

RATE MY UNDERSTANDING
Shade the face that shows your rating

9.11 Thinking about my learning

1 Think back over your learning. Make a list of 10 of the most important and interesting things you learnt on ecosystems. Use this list to frame six questions. Use the 'who, what, where, when, why and how' framework. Clearly write these six questions on the front of an envelope.

2 Rewrite each question on a sheet of paper, allowing room for answers. In a different colour, write detailed answers to each question. Place the answer sheet inside the envelope and seal the envelope.

3 Swap your envelope with a member of your class. Read your partner's questions and answer them in your notebook. Check your answers by opening the envelope and reading the answers supplied.

CHAPTER 10

Plate tectonics

10.1 Knowledge preview

Science understanding

FOUNDATION | **STANDARD** | ADVANCED

1 Tick the box next to the correct definition of 'metamorphic rock':

A rock formed under extremely high temperature and force ☐

B rock formed by grains or sediments ☐

C rock formed when magma cools. ☐

2 The diagram below shows the structure of the Earth. Use the words in the box below to correctly label the different layers of the Earth.

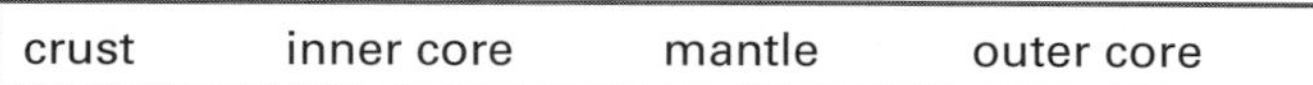

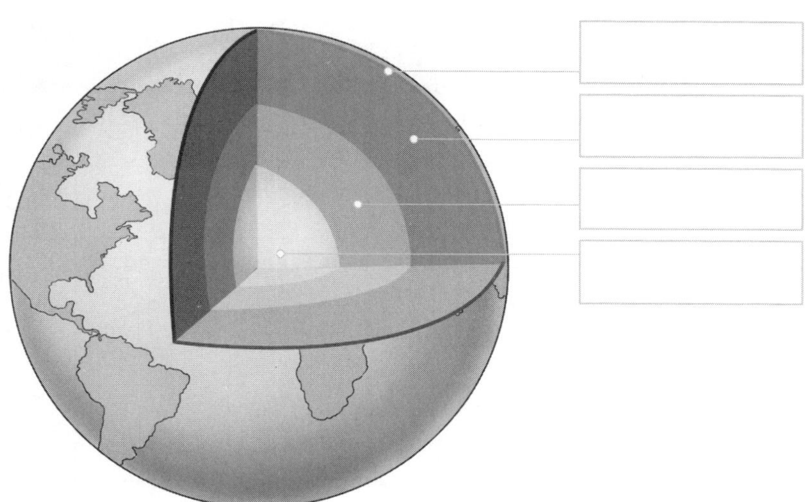

(3) Compare intrusive and extrusive igneous rocks. Write their different and common characteristics on the Venn diagram below.

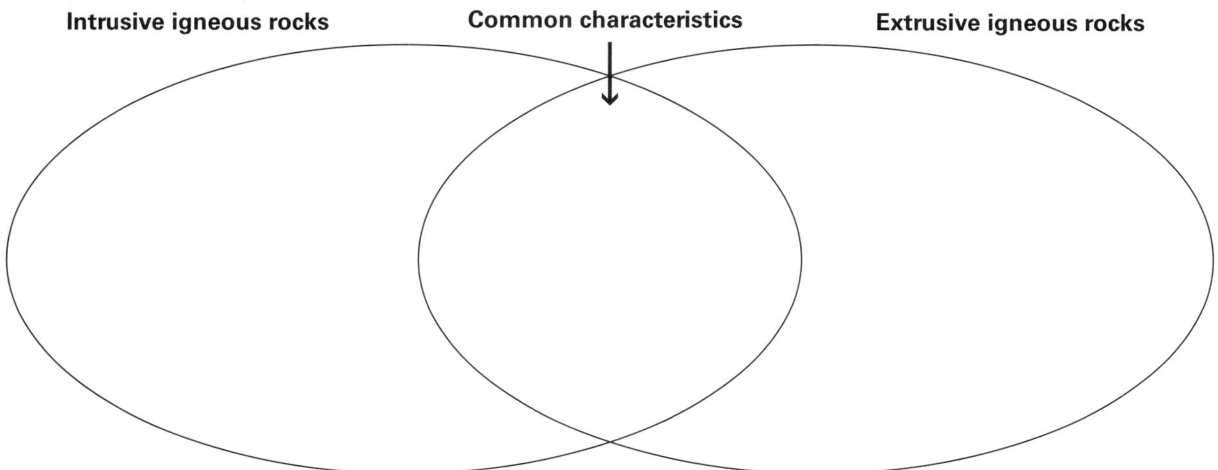

10.1 Knowledge preview

4 Explain why Australia occasionally experiences minor tremors whereas New Zealand, which is only about 2000 km from Australia's east coast, experiences many damaging earthquakes.

5 Look at the list of terms in the box below. Use red to highlight the terms associated with earthquakes and blue to highlight terms associated with volcanoes.

crater	energy waves	epicentre	eruption	focus
hot spot	lava	magma	Richter scale	seismometer

6 Indicate whether the following statements are TRUE or FALSE or if you don't know the answer, circle UNSURE.

Statements	Highlight your answer: TRUE / FALSE / UNSURE
The Earth's continents have always been in the same positions as they are today.	TRUE / FALSE / UNSURE
Millions of years ago Australia was joined to the continent of Antarctica.	TRUE / FALSE / UNSURE
There is no evidence to prove that continents were once all joined together.	TRUE / FALSE / UNSURE
The Atlantic Ocean is slowly getting wider as continental plates separate.	TRUE / FALSE / UNSURE
The edges of continental plates are the locations of many earthquakes and volcanic activity.	TRUE / FALSE / UNSURE
The edge of a continental plate runs through the centre of Australia from north to south.	TRUE / FALSE / UNSURE
Magma and lava are exactly the same thing and the two terms can be used interchangeably.	TRUE / FALSE / UNSURE
Tsunamis may be a result of earthquake activity on the ocean floor.	TRUE / FALSE / UNSURE
Mt Vesuvius, Mt St Helens and Mt Fuji are all volcanic mountains.	TRUE / FALSE / UNSURE
Volcanic and earthquake activity may occur on the Earth's crust on land and under the sea.	TRUE / FALSE / UNSURE

Rebuilding Gondwana

Science understanding

FOUNDATION | **STANDARD** | ADVANCED

Alfred Wegener's theory of continental drift was based on two main observations:

- some continents seemed to fit together a bit like a jigsaw
- fossils of the same species were found in countries that were a long way apart.

Wegener could see no way the animals and plants could cross the oceans to reach all of these places. He joined the continents up so that each species occurred in one continuous strip across the land. His theory made sense because species live in a particular home range rather than scattered widely in separated places.

In this activity, you will try a simplified version of how Wegener joined the continents together. In Figure 10.2.1 you can see the shapes of modern-day continents and islands that Wegener thought belonged together in one large land mass. This land mass was called Gondwana. The different shadings show the distribution of the fossils of four species that lived at the same time. Wegener thought these four fossil species must have lived in the time the land masses were joined together.

fossil (*n*) the remains or impression of a prehistoric plant or animal embedded in rock and preserved

jigsaw puzzle (*n*) a puzzle with a picture cut out into pieces that fit together

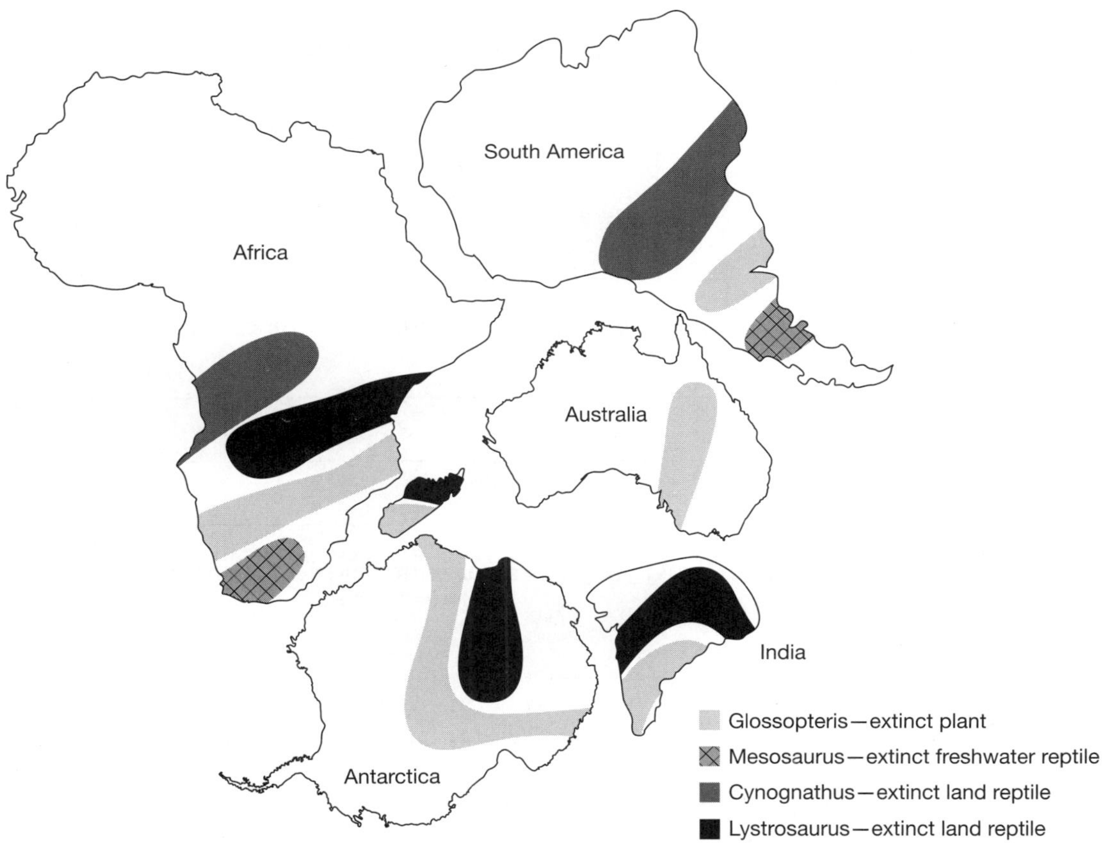

Figure 10.2.1

10.2 Rebuilding Gondwana

1 (a) Copy and cut out the land masses from the previous page (page 143).

(b) Use the fossil distribution and shapes of each land mass to rearrange the continents and islands into one large land mass.

(c) Glue the land mass into the space below. (If necessary, your land masses can be glued over the words here in question 1.)

Glue your Gondwana land mass here.

2 How did the shapes of the land masses and fossil distribution help you create your Gondwana land mass?

____

____

____

3 Explain why Wegener thought that the continents of today were joined together in the past.

____

____

____

4 Based on your findings in this activity, do you agree with Wegener's theory? Justify your answer.

____

____

____

5 Evaluate the importance of Wegener's contribution to science.

____

____

____

RATE MY UNDERSTANDING
Shade the face that shows your rating

10.3 Earth's magnetic field

Science understanding

FOUNDATION	STANDARD	ADVANCED

Earth has a magnetic field around it, as shown in Figure 10.3.1. Scientists have concluded that this field is likely to be caused by the flow of iron and nickel in the liquid part of the Earth's core. As the liquid moves, it creates the magnetic field round it.

Geophysicists have studied rocks of different ages from around Earth that contain magnetite particles. These are like tiny magnets that are trapped in the rock. These magnetic particles would normally point in one direction—towards the north. Studies of these rocks have shown that at times, in rocks of particular ages, the magnetite particles were pointing in the opposite direction, towards the south. From this, scientists concluded that Earth's magnetic field has reversed its orientation many times. A compass in this reversed field would point the opposite way.

Geophysicists have used radioactive dating methods and the magnetic records in the rocks to construct a time strip. This time strip shows when the rocks had normal or reversed magnetism (polarity). A small part of the time strip, called the Geomagnetic Polarity Timescale (GPTS), is shown in Figure 10.3.2. The dates on the left are in millions of years ago (Ma). Chron refers to a major time period, while subchron is a subsection of the major time period. The dates on the subchron side are in millions of years. The time scale was constructed using information from rocks of the ocean floor and from land.

The magnetic field is becoming weaker at a rate that will probably cause it to begin to reverse in about 1000–2000 years from now. Some scientists have proposed that this reduced magnetic field strength will expose Earth to more dangerous cosmic radiation and could lead to extinction of some species. However, other early human ancestors must have survived previous reversals because the last one was 780 000 years ago.

At present Earth's magnetic north is shifting from northern Canada towards Siberia. It is currently moving at about 40 km/year, four times faster than it was moving at the start of the 20th century.

Homo erectus (*n*) an early species of upright man
magnet (*n*) a piece of metal that can attract iron, nickel or cobalt
orientation (*n*) position or direction
subsection (*n*) smaller part of a larger section

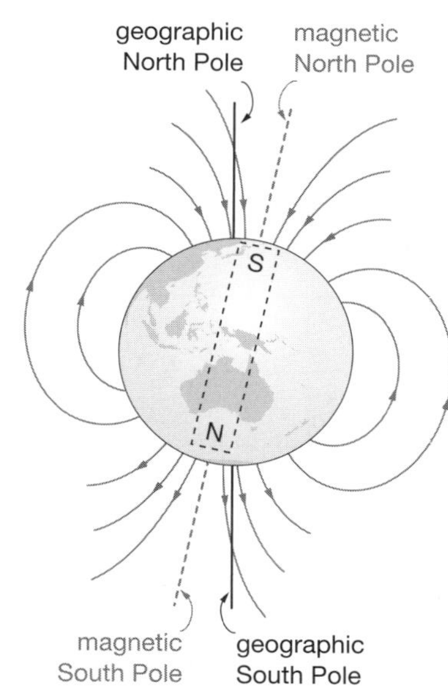

Figure 10.3.1 The Earth's magnetic field. In order for the north end of a compass needle to point towards the north pole, an imaginary bar magnet buried inside the Earth would have its south end at the north pole (opposite poles attract).

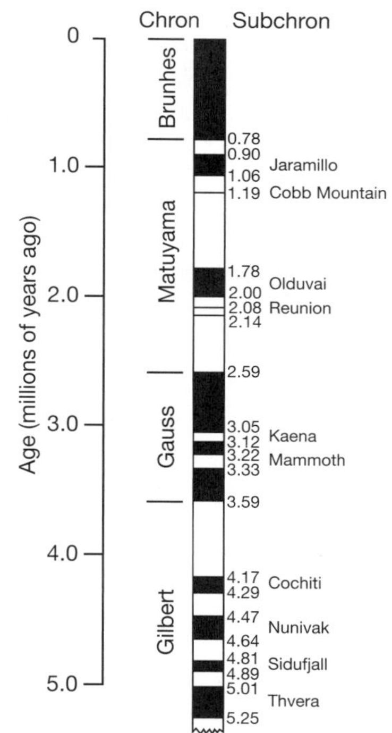

Figure 10.3.2 The GPTS strip for the past 5 million years. The dark areas are normal polarity and the white areas are reversed polarity. The names identify the places where the rocks were found that were used to measure the magnetic field direction.

10.3 Earth's magnetic field

1 Explain the likely reason for Earth having a magnetic field.

(2) Explain the importance of magnetite particles to the discovery of the reversal of Earth's magnetic field.

(3) Describe how the GPTS scale was created.

(4) Use the GPTS strip to deduce how many times Earth's magnetic field has reversed in the past 5 million years.

(5) *Homo erectus* was an early member of the human family. This species was discovered from fossils that date back to Africa about 1.9 million years ago. The youngest fossils of this species date back to about 70 000 years ago. How many times did the magnetic field change while *Homo erectus* existed?

RATE MY UNDERSTANDING
Shade the face that shows your rating   

10.4 Tectonic plates and landforms

Science understanding

FOUNDATION | **STANDARD** | ADVANCED

The Earth's outer shell or crust is made up of a number of large and smaller sections called tectonic plates. These plates move very slowly over the mantle. As plates move, they may collide, slide past each other or move apart. Over long periods of time, plate movements shape the Earth's landscapes. In Figure 10.4.1, the world map shows the major plates, and arrows indicate the general directions in which each plate is moving.

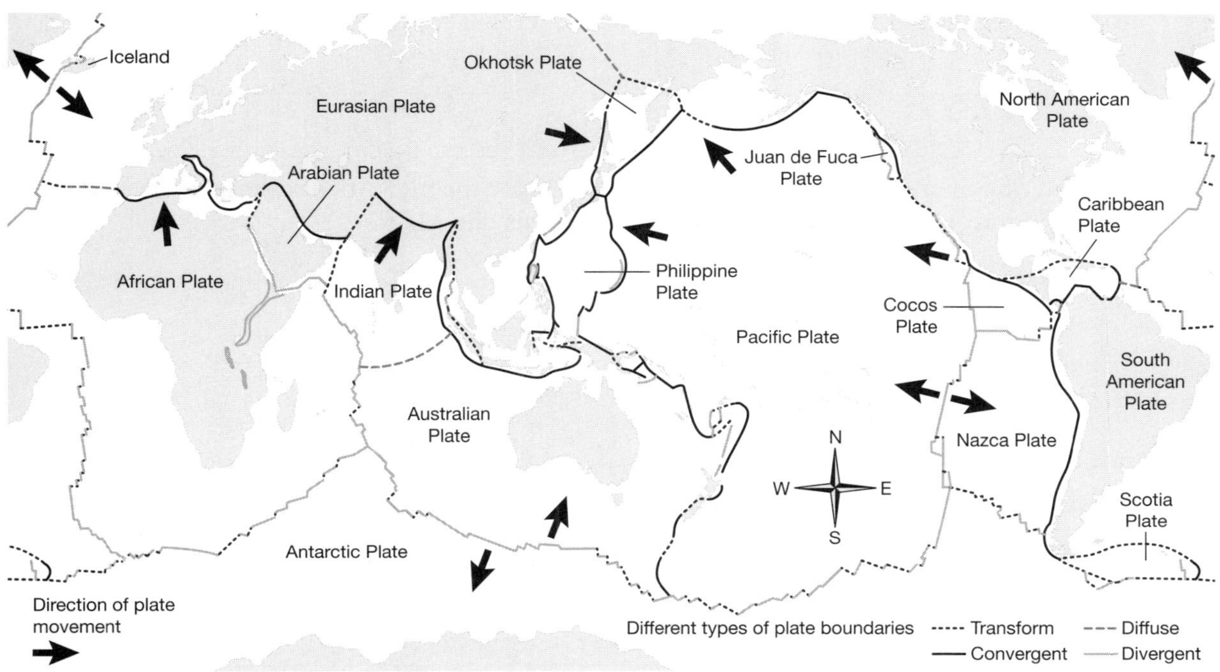

Figure 10.4.1 World map of tectonic plates

Carefully examine each of the diagrams that follow. Each diagram demonstrates a particular movement of plates and the resulting effect on the landscape. Conduct research to help you complete this worksheet if required.

1. Consider Figure 10.4.2. These two diagrams show different parts of the Earth where the plates are moving in a similar way.

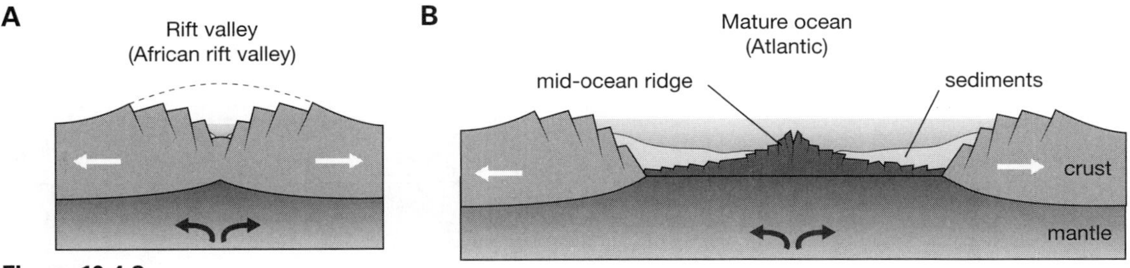

Figure 10.4.2

(a) Refer to an atlas, mapping website or app. Locate the African Rift Valley and write ARV on Figure 10.4.1 to show its location.

(b) The African Rift Valley is an example of a newly developing plate boundary. Describe the plate movement at this location.

(c) Describe what this area may have looked like before the tectonic plates began separating.

(d) Describe the landscape features at this location that result from the plate movement.

(e) Refer to an atlas, mapping website or app. Locate the mid-Atlantic ridge and draw a line on Figure 10.4.1 to show its location. Label this line MAR.

(f) Name the two plates on either side of the ridge.

(g) Describe the plate movement at this location.

(h) Describe what the Atlantic Ocean would have looked like before tectonic plate movement began.

(i) Describe what landscape features are now evident along the mid-Atlantic Ridge, as a result of the plate movement.

2 Use the diagrams in Figure 10.4.2 to explain how new oceans form.

10.4 Tectonic plates and landforms

3 Use Figure 10.4.3 to complete the questions on how island arcs form. This is the type of plate movement along the Indonesian islands and Japan. This is a different plate movement to that in Figure 10.4.2.

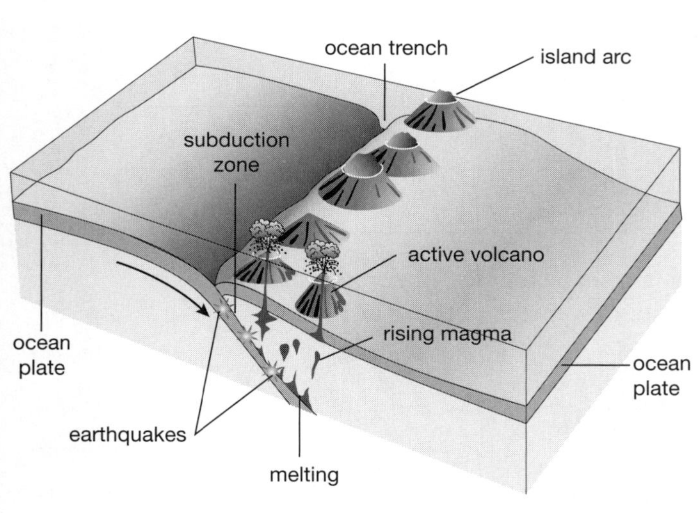

Figure 10.4.3

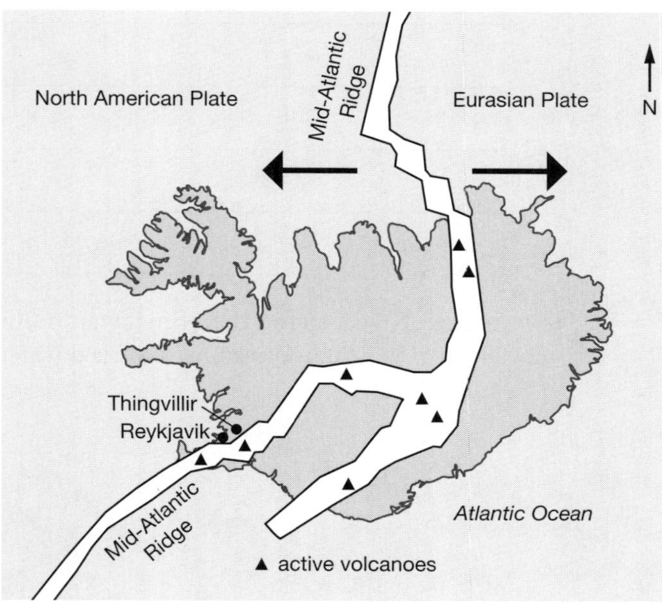

Figure 10.4.4 Iceland and plate movement

(a) Refer to an atlas or mapping website or app. Locate Japan and Indonesia on Figure 10.4.1. Show the location of Japan with a dot and label it J. Indonesia is a bigger country so circle it and label it with IND.

(b) Name the two plates involved in forming islands in Japan.

____

(c) Name the two plates involved in forming islands in Indonesia.

____

(d) Describe the plate movement in either Japan or Indonesia.

____

____

(e) Name some landscape features that have developed as a result of this type of plate movement.

____

____

4 Use Figure 10.4.4 and your knowledge of plate tectonics to complete the following.

(a) Refer to an atlas or mapping website or app. Locate Iceland and place a dot on Figure 10.4.1 to show its location. Label this dot I.

(b) Explain why Iceland can be a dangerous place to live.

____

____

____

(c) What do you predict will change in Iceland's landscape in a million years, if the plates continue to move as they are today? Give reasons for your answer.

5 Figure 10.4.5 shows the Himalayan mountains landscape and the plate movement that has formed the mountains. This is a different plate movement to those in Figures 10.4.2 and 10.4.3.

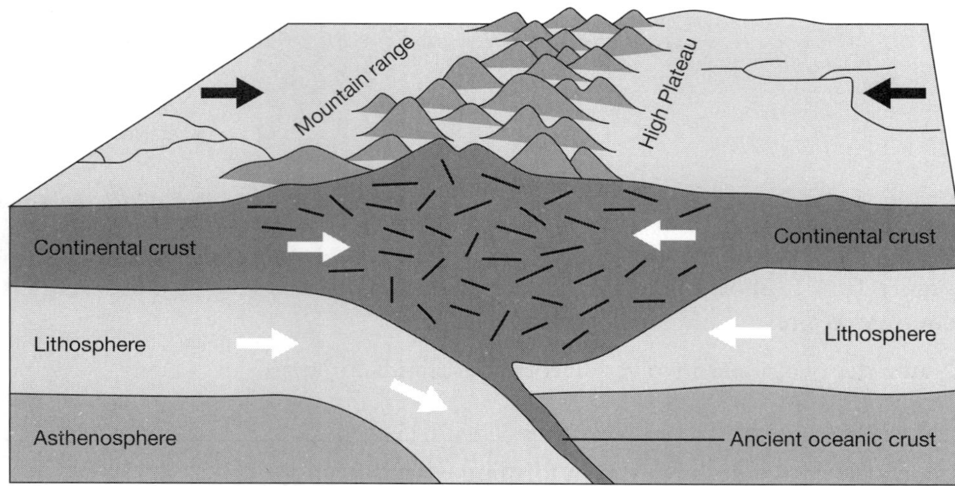

Figure 10.4.5

(a) Refer to an atlas, mapping website or app. Locate the Himalayan mountains and include a line on Figure 10.4.1 to show their location. Label this line H.

(b) Name the two plates involved in forming the Himalayas.

(c) Describe the plate movement at this location.

(d) What effect can colliding plates have on the landscape?

RATE MY UNDERSTANDING
Shade the face that shows your rating  

10.5 Tsunami

Science understanding

FOUNDATION | **STANDARD** | ADVANCED

On 26 December 2004, a tsunami hit the islands off the western coast of the Indonesian island of Sumatra. The tsunami killed about 230 000–280 000 people, mainly in Indonesia, Sri Lanka, India and Thailand. The tsunami was caused by a sudden slip of the tectonic plate boundary between the Indo-Australian Plate and a section of the Eurasian Plate called the Sunda Plate. Between these plates is a deep trench, called the Sunda Trench. You can see its location in Figure 10.5.1.

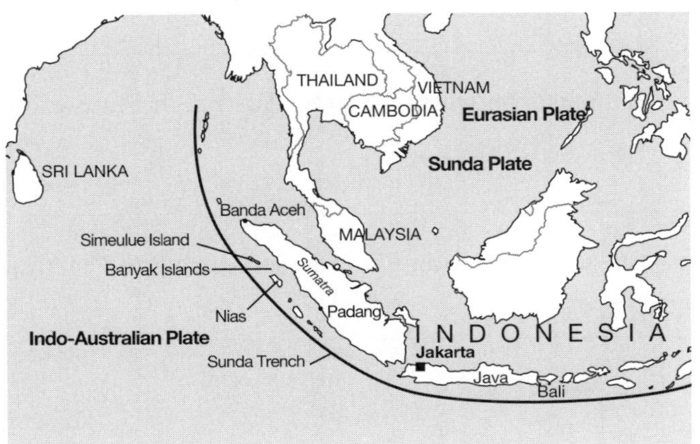

Figure 10.5.1 The Sunda Plate in South-East Asia

ebb (*n*) the movement of the tide away from the land

epicentre (*n*) the point on the Earth's surface above the point where an earthquake starts

crest (*n*) the highest point of a wave

trench (*n*) a valley

trough (*n*) the lowest point of a wave

As the fault line slipped, it triggered a massive earthquake of magnitude 9.0 on the Richter scale. The epicentre was under the sea near Simeulue Island, and large waves about 37 metres high formed at the surface. The sequence of events is shown in Figure 10.5.2.

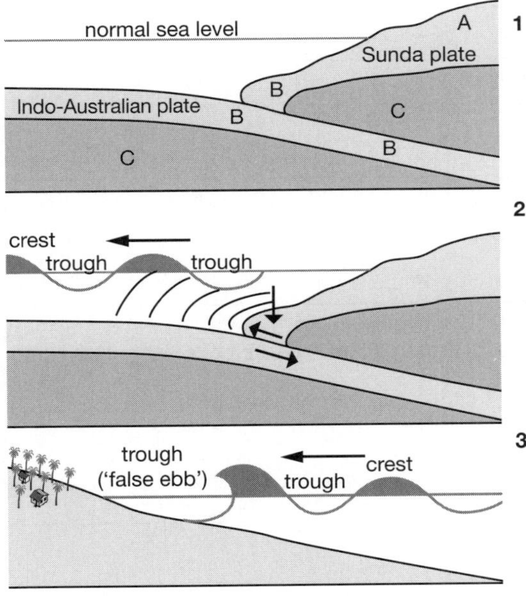

1 Before the earthquake
A Continental crust
B Oceanic crust and lithosphere
C Asthenosphere

2 Formation of the tsunami. The Indo-Australian Plate slips down by 20 m and the Sunda Plate is uplifted by 5 m. A tsunami travels outwards initially at 700 km/h in the deep ocean, while a surface wave 37 m high forms.

3 Tsunami waves hitting land. Tsunamis slow down in shallow water. The wave may become much higher, depending on the landscape, and the crests of the waves are closer together. Just before the wave hits the beach, the water level can drop as if the tide has gone out. This is called a false ebb.

Figure 10.5.2 Tsunami formation

1 What is a tsunami?

___

___

10.5 Tsunami

2 What caused the tsunami of 26 December 2004?

3 What is an epicentre?

(4) Identify the type of plate boundary that exists between the Indo-Australian Plate and the Sunda Plate.

5 State how far the Indo-Australian Plate and the Sunda Plate moved during the earthquake.

6 What is the speed of a tsunami in deep water?

(7) What effect does a tsunami have when it approaches and enters shallow water?

8 List warning signs that a tsunami could be approaching a beach.

(9) Australia donated about $1377 million to Indonesia to aid their recovery. Propose reasons why.

RATE MY UNDERSTANDING
Shade the face that shows your rating

10.6 Plate tectonics summary

Science understanding

FOUNDATION | **STANDARD** | ADVANCED

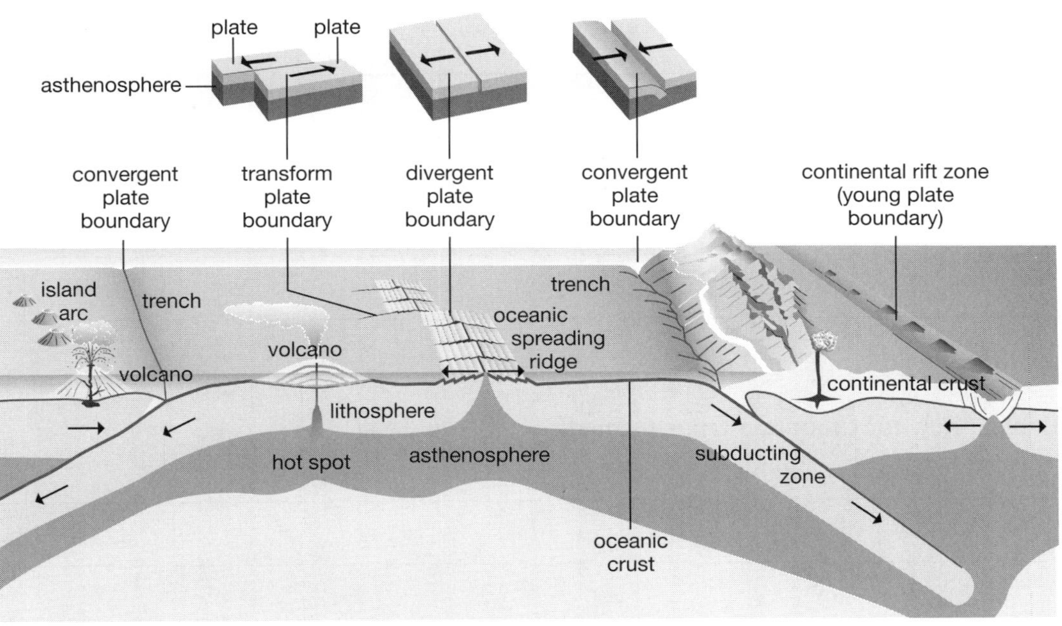

Figure 10.6.1

1 Where is seafloor spreading shown in Figure 10.6.1?

______

______

______

(2) Suggest why transform faults form near ocean ridges.

______

______

______

(3) What do you think might happen in the future on the plate on the right side of Figure 10.6.1 labelled continental rift zone?

______

______

______

10.6 Plate tectonics summary

4. Explain what the subducting zone refers to.

5. Explain how the island arc in Figure 10.6.1 formed.

6. Explain how the mountain range formed.

7. Explain how the trenches formed.

RATE MY UNDERSTANDING
Shade the face that shows your rating

10.7 Literacy review

Science understanding

FOUNDATION	STANDARD	ADVANCED

1 Complete the following crossword on earthquakes and plate tectonics. Do this by reading the clues and writing the correct words into the crossword and creating hints for words already in the crossword.

															1 A				
											2				S				
															T				
											7 S				H				
											E				E				
											I				N				
											S				O				
											M				S				3
							4				O		5		P				
									6		M				H				
							8 E	P	I	C	E	N	T	R	E				
											T				R				
											E				E				
											R								
													9 T	S	U	N	A	M	I
10																			

ACROSS

2 The place where the earthquake starts below ground is called the ________ .

8 ______________________

9 ______________________

10 A boundary where plates are colliding is known as a __________ boundary.

DOWN

1 ______________________

3 When one plate is forced below another plate it is called ______________ .

4 A channel in the ocean floor formed where plates collide and the crust subducts is called a __________ .

5 A ________ is an isolated place away from plate boundaries where a lot of hot magma is created.

6 ___________ is the process of continents breaking up.

7 ______________________________

2 In the space provided, write the correct term for each description.

(a) The crust and the solid part of the upper mantle that together make up a tectonic plate ______________ .

(b) Process of continents separating and drifting across the Earth ______________ .

(c) The theory that the Earth's crust is cracked into large pieces that slowly move position ______________ .

(d) A chain of islands formed at the edge of colliding tectonic plates where one plate subducts ______________ .

(e) Longitudinal seismic waves that travel fast through the Earth ______________ .

(f) Transverse seismic waves that travel through the Earth ____________________ .

(g) Patterns of magnetism trapped in rocks on each side of diverging plate boundaries ____________________ .

(h) The process of new crust forming at the ocean ridges and spreading outwards ____________________ .

(i) Type of boundary where tectonic plates are sliding parallel to each other but in opposite directions ____________________ .

RATE MY UNDERSTANDING
Shade the face that shows your rating  

10.8 Thinking about my learning

How does your thinking shape up? How do you feel about how you have done in understanding this chapter? Complete the statements in the shapes below. Use the hints to help you.

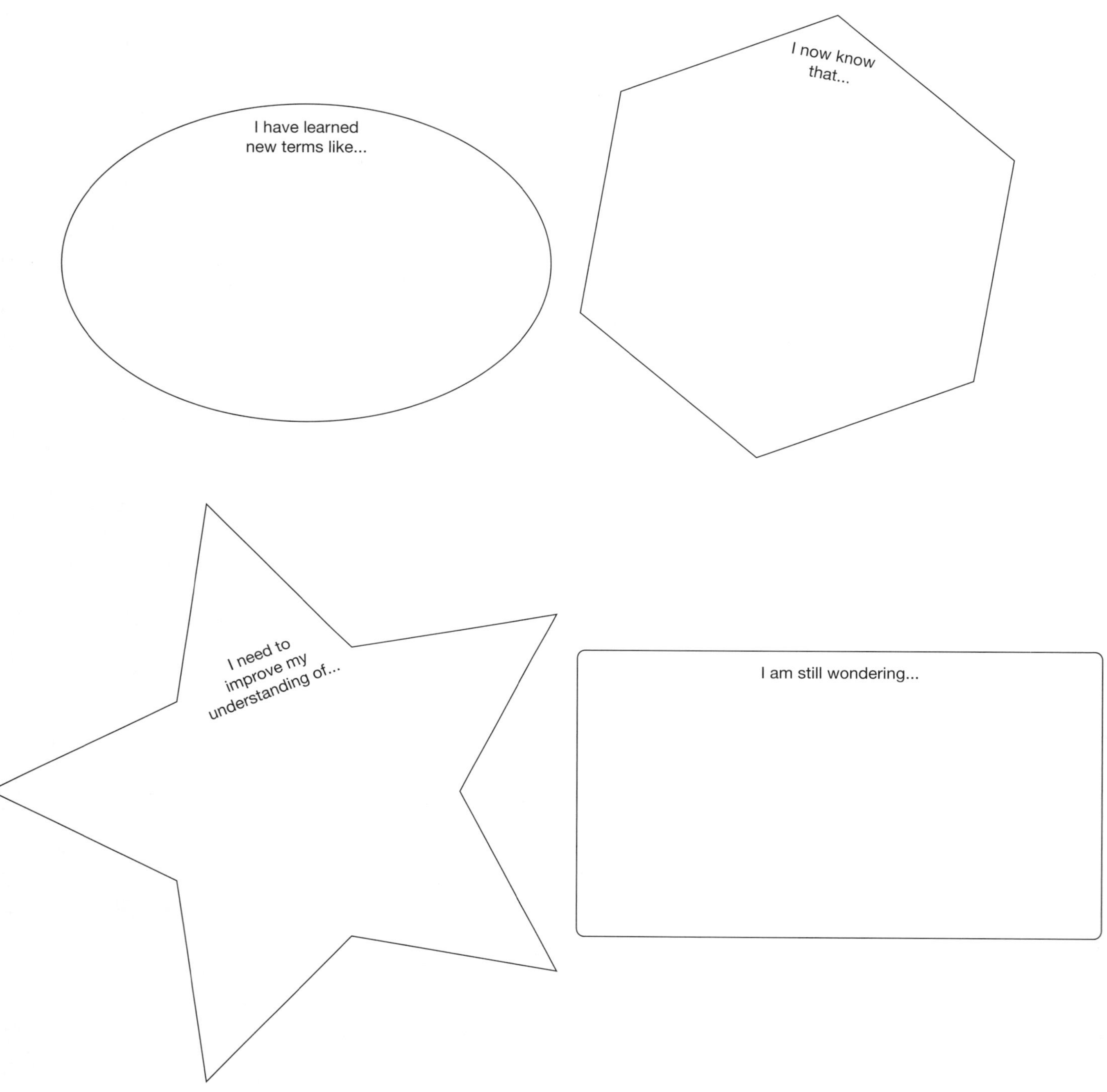